PRINCIPLES OF MINING

VALUATION, ORGANIZATION AND ADMINISTRATION COPPER, GOLD, LEAD, SILVER, TIN AND ZINC

BY

HERBERT C. HOOVER

Member American Institute of Mining Engineers, Mining and Metallurgical Society of America, Société des Ingénieurs Civils de France, Fellow Royal Geographical Society, etc.

PREFACE.

This volume is a condensation of a series of lectures delivered in part at Stanford and in part at Columbia Universities. It is intended neither for those wholly ignorant of mining, nor for those long experienced in the profession.

The bulk of the material presented is the common heritage of the profession, and if any one may think there is insufficient reference to previous writers, let him endeavor to find to whom the origin of our methods should be credited. The science has grown by small contributions of experience since, or before, those unnamed Egyptian engineers, whose works prove their knowledge of many fundamentals of mine engineering six thousand eight hundred years ago. If I have contributed one sentence to the accumulated knowledge of a thousand generations of engineers, or have thrown one new ray of light on the work, I shall have done my share.

I therefore must acknowledge my obligations to all those who have gone before, to all that has been written that I have read, to those engineers with whom I have been associated for many years, and in particular to many friends for kindly reply to inquiry upon points herein discussed.

CONTENTS.

ENTRY TO THE MINE; TUNNELS; VERTICAL, INCLINED, AND COMBINED SHAFTS; LOCATION AND NUMBER OF SHAFTS.

PRINCIPLES OF MINING.

CHAPTER I.

VALUATION OF COPPER, GOLD, LEAD, SILVER, TIN, AND ZINC LODE MINES.

DETERMINATION OF AVERAGE METAL CONTENT; SAMPLING, ASSAY PLANS, CALCULATIONS OF AVERAGES, PERCENTAGE OF ERRORS IN ESTIMATE FROM SAMPLING.

The following discussion is limited to *in situ* deposits of copper, gold, lead, silver, tin, and zinc. The valuation of alluvial deposits, iron, coal, and other mines is each a special science to itself and cannot be adequately discussed in common with the type of deposits mentioned above.

The value of a metal mine of the order under discussion depends upon:—

 a. The profit that may be won from ore exposed;
 b. The prospective profit to be derived from extension of the ore beyond exposures;
 c. The effect of a higher or lower price of metal (except in gold mines);
 d. The efficiency of the management during realization.

The first may be termed the positive value, and can be approximately determined by sampling or test-treatment runs. The second and the third may be termed the speculative values, and are largely a matter of judgment based on geological evidence and the industrial outlook. The fourth is a question of development, equipment, and engineering method adapted to the prospects of the enterprise, together with capable executive control of these works.

It should be stated at the outset that it is utterly impossible to accurately value any mine, owing to the many speculative factors involved. The best that can be done is to state that the value lies between certain limits, and that various stages above the minimum given represent various degrees of risk. Further, it would be but stating truisms to those engaged in valuing mines to repeat that, because of the limited life of every mine, valuation of such investments cannot be based upon the principle of simple interest; nor that any investment is justified without a consideration of the management to ensue. Yet the ignorance of these essentials is so prevalent among the public that they warrant repetition on every available occasion.

To such an extent is the realization of profits indicated from the other factors dependent upon the subsequent management of the enterprise that the author considers a review of underground engineering and administration from an economic point of view an essential to any essay upon the subject. While the metallurgical treatment of ores is an essential factor in mine economics, it is considered that a detailed discussion of the myriad of processes under hypothetic conditions would lead too far afield. Therefore the discussion is largely limited to underground and administrative matters.

The valuation of mines arises not only from their change of ownership, but from the necessity in sound administration for a knowledge of some of the fundamentals of valuation, such as ore reserves and average values, that managerial and financial policy may be guided aright. Also with the growth of corporate ownership there is a demand from owners and stockholders for periodic information as to the intrinsic condition of their properties.

The growth of a body of speculators and investors in mining stocks and securities who desire professional guidance which cannot be based upon first-hand data is creating further demand on the engineer. Opinions in these cases must be formed on casual visits or second-hand information, and a knowledge of men and things generally. Despite the feeling of some engineers that the latter employment is not properly based professionally, it is an expanding phase of engineers' work, and must be taken seriously. Although it lacks satisfactory foundation for accurate judgment, yet the engineer can, and should, give his experience to it when the call comes, out of interest to the industry as a whole. Not only can he in a measure protect the lamb, by insistence on no investment without the provision of properly organized data and sound administration for his client, but he can do much to direct the industry from gambling into industrial lines.

An examination of the factors which arise on the valuation of mines involves a wide range of subjects. For purposes of this discussion they may be divided into the following heads:—

1. *Determination of Average Metal Contents of the Ore.*
2. *Determination of Quantities of Ore.*
3. *Prospective Value.*
4. *Recoverable Percentage of Gross Value.*
5. *Price of Metals.*
6. *Cost of Production.*
7. *Redemption or Amortization of Capital and Interest.*
8. *Valuation of Mines without Ore in Sight.*
9. *General Conduct of Examination and Reports.*

DETERMINATION OF AVERAGE METAL CONTENTS OF THE ORE.

Three means of determination of the average metal content of standing ore are in use—Previous Yield, Test-treatment Runs, and Sampling.

Previous Yield.—There are certain types of ore where the previous yield from known space becomes the essential basis of determination of quantity and metal contents of ore standing and of the future probabilities. Where metals occur like plums in a pudding, sampling becomes difficult and unreliable, and where experience has proved a sort of regularity of recurrence of these plums, dependence must necessarily be placed on past records, for if their reliability is to be questioned, resort must be had to extensive test-treatment runs. The Lake Superior copper mines and the Missouri lead and zinc mines are of this type of deposit. On the other sorts of deposits the previous yield is often put forward as of important bearing on the value of the ore standing, but such yield, unless it can be *authentically* connected with blocks of ore remaining, is not necessarily a criterion of their contents. Except in the cases mentioned, and as a check on other methods of determination, it has little place in final conclusions.

Test Parcels.—Treatment on a considerable scale of sufficiently regulated parcels, although theoretically the ideal method, is, however, not often within the realm of things practical. In examination on behalf of intending purchasers, the time, expense, or opportunity to fraud are usually prohibitive, even where the plant and facilities for such work exist. Even in cases where the engineer in management of producing mines is desirous of determining the value of standing ore, with the exception of deposits of the type mentioned above, it is ordinarily done by actual sampling, because separate mining and treatment of test lots is generally inconvenient and expensive. As a result, the determination of the value of standing ore is, in the great majority of cases, done by sampling and assaying.

Sampling.—The whole theory of sampling is based on the distribution of metals through the ore-body with more or less regularity, so that if small portions, that is samples, be taken from a sufficient number of points, their average will represent fairly closely the unit value of the ore. If the ore is of the extreme type of irregular metal distribution mentioned under "previous yield," then sampling has no place.

How frequently samples must be taken, the manner of taking them, and the quantity that constitutes a fair sample, are matters that vary with each mine. So much depends upon the proper performance of this task that it is in fact the most critical feature of mine examination. Ten samples properly taken are more valuable than five hundred slovenly ones, like grab samples, for such a number of bad ones would of a surety lead to wholly wrong conclusions. Given a good sampling and a proper assay plan, the valuation of a mine is two-thirds accomplished. It should be an inflexible principle in examinations for purchase that every sample must be taken under the personal supervision of the examining engineer or his trusted assistants. Aside from throwing open the doors to fraud, the average work-man will not carry out the work in a proper manner, unless under constant supervision, because of his lack of appreciation of the issues involved. Sampling is hard, uncongenial, manual labor. It requires a deal of conscientiousness to take enough samples and to take them thoroughly. The engineer does not exist who, upon completion of this task, considers that he has got too many, and most wish that they had taken more.

The accuracy of sampling as a method of determining the val-ue of standing ore is a factor of the number of samples taken. The average, for example, of separate samples from each square inch would be more accurate than those from each alternate square inch. However, the accumulated knowledge and expe-rience as to the distribution of metals through ore has determined approximately the manner of taking such samples, and the least number which will still by the law of averages secure a degree of accuracy commensurate with the other factors of estimation.

As metals are distributed through ore-bodies of fissure origin with most regularity on lines parallel to the strike and dip, an equal portion of ore from every point along cross-sections at right angles to the strike will represent fairly well the average values for a certain distance along the strike either side of these cross-sections. In massive deposits, sample sections are taken in all directions. The intervals at which sample sections must be cut is obviously dependent upon the general character of the deposit. If the values are well distributed, a longer interval may be em-ployed than in one subject to marked fluctuations. As a general rule, five feet is the distance most accepted. This, in cases of regular distribution of values, may be stretched to ten feet, or in reverse may be diminished to two or three feet.

The width of ore which may be included for one sample is de-pendent not only upon the width of the deposit, but also upon its character. Where the ore is wider than the necessary stoping width, the sample should be regulated so as to show the possible locus of values. The metal contents may be, and often are, particularly in deposits of the impregnation or replacement type, greater along some streak in the ore-body, and this difference may be such as to make it desirable to stope only a portion of the total thickness. For deposits narrower than the necessary stoping width the full breadth of ore should be included in one sample, because usually the whole of the deposit will require to be broken.

In order that a payable section may not possibly be diluted with material unnecessary to mine, if the deposit is over four feet and under eight feet, the distance across the vein or lode is usually divided into two samples. If still wider, each is confined to a span of about four feet, not only for the reason given above, but because the more numerous the samples, the greater the accuracy. Thus, in a deposit twenty feet wide it may be taken as a good guide that a test section across the ore-body should be divided into five parts.

As to the physical details of sample taking, every engineer has his own methods and safeguards against fraud and error. In a large organization of which the writer had for some years the direction, and where sampling of mines was constantly in progress on an extensive scale, not only in contemplation of purchase, but where it was also systematically conducted in operating mines for working data, he adopted the above general lines and required the following details.

A fresh face of ore is first broken and then a trench cut about five inches wide and two inches deep. This trench is cut with a hammer and moil, or, where compressed air is available and the rock hard, a small air-drill of the hammer type is used. The spoil from the trench forms the sample, and it is broken down upon a large canvas cloth. Afterwards it is crushed so that all pieces will pass a half-inch screen, mixed and quartered, thus reducing the weight to half. Whether it is again crushed and quartered depends upon what the conditions are as to assaying. If convenient to assay office, as on a going mine, the whole of the crushing and quartering work can be done at that office, where there are usually suitable mechanical appliances. If the samples must be taken a long distance, the bulk for transport can be reduced by finer breaking and repeated quartering, until there remain only a few ounces.

Precautions against Fraud.—Much has been written about the precautions to be taken against fraud in cases of valuations for purchase. The best safeguards are an alert eye and a strong right arm. However, certain small details help. A large leather bag, arranged to lock after the order of a mail sack, into which samples can be put underground and which is never unfastened except by responsible men, not only aids security but relieves the mind. A few samples of country rock form a good check, and notes as to the probable value of the ore, from inspection when sampling, are useful. A great help in examination is to have the assays or analyses done coincidentally with the sampling. A doubt can then always be settled by resampling at once, and much knowledge can be gained which may relieve so exhaustive a program as might be necessary were results not known until after leaving the mine.

Assay of Samples.—Two assays, or as the case may be, analyses, are usually made of every sample and their average taken. In the case of erratic differences a third determination is necessary.

Assay Plans.—An assay plan is a plan of the workings, with the location, assay value, and width of the sample entered upon it. In a mine with a narrow vein or ore-body, a longitudinal section is sufficient base for such entries, but with a greater width than one sample span it is desirable to make preliminary plans of separate levels, winzes, etc., and to average the value of the whole payable widths on such plans before entry upon a longitudinal section. Such a longitudinal section will, through the indicated distribution of values, show the shape of the ore-body—a step necessary in estimating quantities and of the most fundamental importance in estimating the probabilities of ore extension beyond the range of the openings. The final assay plan should show the average value of the several blocks of ore, and it is from these averages that estimates of quantities must be made up.

Calculations of Averages.—The first step in arriving at average values is to reduce erratic high assays to the general tenor of other adjacent samples. This point has been disputed at some length, more often by promoters than by engineers, but the custom is very generally and rightly adopted. Erratically high samples may indicate presence of undue metal in the assay attributable to unconscious salting, for if the value be confined to a few large particles they may find their way through all the quartering into the assay. Or the sample may actually indicate rich spots of ore; but in any event experience teaches that no dependence can be put upon regular recurrence of such abnormally rich spots. As will be discussed under percentage of error in sampling, samples usually indicate higher than the true value, even where erratic assays have been eliminated. There are cases of profitable mines where the values were all in spots, and an assay plan would show 80% of the assays *nil*, yet these pockets were so rich as to give value to the whole. Pocket mines, as stated before, are beyond valuation by sampling, and aside from the previous yield recourse must be had to actual treatment runs on every block of ore separately.

After reduction of erratic assays, a preliminary study of the runs of value or shapes of the ore-bodies is necessary before any calculation of averages. A preliminary delineation of the boundaries of the payable areas on the assay plan will indicate the sections of the mine which are unpayable, and from which therefore samples can be rightly excluded in arriving at an average of the payable ore (Fig. 1). In a general way, only the ore which must be mined need be included in averaging.

The calculation of the average assay value of standing ore from samples is one which seems to require some statement of elementals. Although it may seem primitive, it can do no harm to recall that if a dump of two tons of ore assaying twenty ounces per ton be added to a dump of five tons averaging one ounce per ton, the result has not an average assay of twenty-one ounces divided by the number of dumps. Likewise one sample over a width of two feet, assaying twenty ounces per ton, if averaged with another sample over a width of five feet, assaying one ounce, is no more twenty-one ounces divided by two samples than in the case of the two dumps. If common sense were not sufficient demonstration of this, it can be shown algebraically. Were samples equidistant from each other, and were they of equal width, the average value would be the simple arithmetical mean of the assays. But this is seldom the case. The number of instances, not only in practice but also in technical literature, where the fundamental distinction between an arithmetical and a geometrical mean is lost sight of is amazing.

To arrive at the average value of samples, it is necessary, in effect, to reduce them to the actual quantity of the metal and volume of ore represented by each. The method of calculation therefore is one which gives every sample an importance depending upon the metal content of the volume of ore it represents.

The volume of ore appertaining to any given sample can be considered as a prismoid, the dimensions of which may be stated as follows:—

W Width in feet of ore sampled.

L Length in feet of ore represented by the sample.

D Depth into the block to which values are assumed to penetrate.

We may also let:—

C The number of cubic feet per ton of ore.

V Assay value of the sample.

Then $\dfrac{WLD}{C}$ tonnage of the prismoid.[1]

$\dfrac{V}{WLD/C}$ total metal contents.

The average value of a number of samples is the total metal contents of their respective prismoids, divided by the total tonnage of these prismoids. If we let W, W_1, V, V_1 etc., represent different samples, we have:—

---

[1] Strictly, the prismoidal formula should be used, but it complicates the study unduly, and for practical purposes the above may be taken as the volume.

$$\frac{V(WLD/C) + V_1(W_1\,L_1D_1/C) + V_2(W_2\,L_2D_2/C)}{WLD/C + W_1L_1D_1/C + W_2L_2D_2/C} = \text{average value.}$$

This may be reduced to:—

$$\frac{(VWLD) + (V_1\,W_1\,L_1D_1) + (V_2\,W_2\,L_2D_2,), \text{ etc.}}{(WLD) + (W_1L_1D_1) + (W_2L_2D_2), \text{ etc.}}$$

As a matter of fact, samples actually represent the value of the outer shell of the block of ore only, and the continuity of the same values through the block is a geological assumption. From the outer shell, all the values can be taken to penetrate equal distances into the block, and therefore D, D_1,D_2 may be considered as equal and the equation becomes:—

$$\frac{(VWL) + (V_1W_1L_1) + (V_2W_2L_2), \text{ etc.}}{(WL) + (W_1L_1) + (W_2L_2), \text{ etc.}}$$

The length of the prismoid base L for any given sample will be a distance equal to one-half the sum of the distances to the two adjacent samples. As a matter of practice, samples are usually taken at regular intervals, and the lengths L, L_1,L_2 becoming thus equal can in such case be eliminated, and the equation becomes:—

$$\frac{(VW) + (V_1W_1) + (V_2W_2), \text{ etc.}}{W + W_1 + W_2, \text{ etc.}}$$

The name "assay foot" or "foot value" has been given to the relation VW, that is, the assay value multiplied by the width sampled.[1] It is by this method that all samples must be averaged. The same relation obviously can be evolved by using an inch instead of a foot, and in narrow veins the assay inch is generally used.

Where the payable cross-section is divided into more than one sample, the different samples in the section must be averaged by

---

[1] An error will be found in this method unless the two end samples be halved, but in a long run of samples this may be disregarded.

the above formula, before being combined with the adjacent section. Where the width sampled is narrower than the necessary stoping width, and where the waste cannot be broken separately, the sample value must be diluted to a stoping width. To dilute narrow samples to a stoping width, a blank value over the extra width which it is necessary to include must be averaged with the sample from the ore on the above formula. Cases arise where, although a certain width of waste must be broken with the ore, it subsequently can be partially sorted out. Practically nothing but experience on the deposit itself will determine how far this will restore the value of the ore to the average of the payable seam. In any event, no sorting can eliminate all such waste; and it is necessary to calculate the value on the breaking width, and then deduct from the gross tonnage to be broken a percentage from sorting. There is always an allowance to be made in sorting for a loss of good ore with the discards.

Percentage of Error in Estimates from Sampling.—It must be remembered that the whole theory of estimation by sampling is founded upon certain assumptions as to evenness of continuity and transition in value and volume. It is but a basis for an estimate, and an estimate is not a statement of fact. It cannot therefore be too forcibly repeated that an estimate is inherently but an approximation, take what care one may in its founding. While it is possible to refine mathematical calculation of averages to almost any nicety, beyond certain essentials it adds nothing to accuracy and is often misleading.

It is desirable to consider where errors are most likely to creep in, assuming that all fundamental data are both accurately taken and considered. Sampling of ore *in situ* in general has a tendency to give higher average value than the actual reduction of the ore will show. On three West Australian gold mines, in records covering a period of over two years, where sampling was most exhaustive as a daily régime of the mines, the values indicated by sampling were 12% higher than the mill yield plus the contents of the residues. On the Witwatersrand gold mines, the actual extractable value is generally considered to be about 78 to 80% of the average shown by sampling, while the mill extractions are on average about 90 to 92% of the head value coming to the mill. In other words, there is a constant discrepancy of about 10 to 12% between the estimated value as indicated by mine samples, and the actual value as shown by yield plus the residues. At Broken Hill, on three lead mines, the yield is about 12% less than sampling would indicate. This constancy of error in one direction has not been so generally acknowledged as would be desirable, and it must be allowed for in calculating final results. The causes of the exaggeration seem to be:—

First, inability to stope a mine to such fine limitations of width, or exclusion of unpayable patches, as would appear practicable when sampling, that is by the inclusion when mining of a certain amount of barren rock. Even in deposits of about normal stoping width, it is impossible to prevent the breaking of a certain amount of waste, even if the ore occurrence is regularly confined by walls.

If the mine be of the impregnation type, such as those at Goldfield, or Kalgoorlie, with values like plums in a pudding, and the stopes themselves directed more by assays than by any physical differences in the ore, the discrepancy becomes very much increased. In mines where the range of values is narrower than the normal stoping width, some wall rock must be broken. Although it is customary to allow for this in calculating the average value from samples, the allowance seldom seems enough. In mines where the ore is broken on to the top of stopes filled with waste, there is some loss underground through mixture with the filling.

Second, the metal content of ores, especially when in the form of sulphides, is usually more friable than the matrix, and in actual breaking of samples an undue proportion of friable material usually creeps in. This is true more in lead, copper, and zinc, than in gold ores. On several gold mines, however, tests on accumulated samples for their sulphide percentage showed a distinctly greater ratio than the tenor of the ore itself in the mill. As the gold is usually associated with the sulphides, the samples showed higher values than the mill.

In general, some considerable factor of safety must be allowed after arriving at calculated average of samples,—how much it is difficult to say, but, in any event, not less than 10%.

CHAPTER II.

Mine Valuation (*Continued*).

CALCULATION OF QUANTITIES OF ORE, AND
CLASSIFICATION OF ORE IN SIGHT.

As mines are opened by levels, rises, etc., through the ore, an extension of these workings has the effect of dividing it into "blocks." The obvious procedure in determining tonnages is to calculate the volume and value of each block separately. Under the law of averages, the multiplicity of these blocks tends in proportion to their number to compensate the percentage of error which might arise in the sampling or estimating of any particular one. The shapes of these blocks, on longitudinal section, are often not regular geometrical figures. As a matter of practice, however, they can be subdivided into such figures that the total will approximate the whole with sufficient closeness for calculations of their areas.

The average width of the ore in any particular block is the arithmetical mean of the width of the sample sections in it,[1] if the samples be an equal distance apart. If they are not equidistant, the average width is the sum of the areas between samples, divided by the total length sampled. The cubic foot contents of a particular block is obviously the width multiplied by the area of its longitudinal section.

The ratio of cubic feet to tons depends on the specific gravity of the ore, its porosity, and moisture. The variability of ores throughout the mine in all these particulars renders any method of calculation simply an approximation in the end. The factors which must remain unknown necessarily lead the engineer to the

[1] This is not strictly true unless the sum of the widths of the two end-sections be divided by two and the result incorporated in calculating the means. In a long series that error is of little importance.

provision of a margin of safety, which makes mathematical refinement and algebraic formulæ ridiculous.

There are in general three methods of determination of the specific volume of ores:—

First, by finding the true specific gravity of a sufficient number of representative specimens; this, however, would not account for the larger voids in the ore-body and in any event, to be anything like accurate, would be as expensive as sampling and is therefore of little more than academic interest.

Second, by determining the weight of quantities broken from measured spaces. This also would require several tests from different portions of the mine, and, in examinations, is usually inconvenient and difficult. Yet it is necessary in cases of unusual materials, such as leached gossans, and it is desirable to have it done sooner or later in going mines, as a check.

Third, by an approximation based upon a calculation from the specific gravities of the predominant minerals in the ore. Ores are a mixture of many minerals; the proportions vary through the same ore-body. Despite this, a few partial analyses, which are usually available from assays of samples and metallurgical tests, and a general inspection as to the compactness of the ore, give a fairly reliable basis for approximation, especially if a reasonable discount be allowed for safety. In such discount must be reflected regard for the porosity of the ore, and the margin of safety necessary may vary from 10 to 25%. If the ore is of unusual character, as in leached deposits, as said before, resort must be had to the second method.

The following table of the weights per cubic foot and the number of cubic feet per ton of some of the principal ore-forming minerals and gangue rocks will be useful for approximating the weight of a cubic foot of ore by the third method. Weights are in pounds avoirdupois, and two thousand pounds are reckoned to the ton.

	WEIGHT PER CUBIC FOOT	NUMBER OF CUBIC FEET PER TON OF 2000 LB.
Antimony	417.50	4.79
Sulphide	285.00	7.01
Arsenical Pyrites	371.87	5.37
Barium Sulphate	278.12	7.19
Calcium:		
Fluorite	198.75	10.06
Gypsum	145.62	13.73
Calcite	169.37	11.80
Copper	552.50	3.62
Calcopyrite	262.50	7.61
Bornite	321.87	6.21
Malachite	247.50	8.04
Azurite	237.50	8.42
Chrysocolla	132.50	15.09
Iron (Cast)	450.00	4.44
Magnetite	315.62	6.33
Hematite	306.25	6.53
Limonite	237.50	8.42
Pyrite	312.50	6.40
Carbonate	240.62	8.31
Lead	710.62	2.81
Galena	468.75	4.27
Carbonate	406.87	4.81
Manganese Oxide	268.75	6.18
Rhodonite	221.25	9.04
Magnesite	187.50	10.66
Dolomite	178.12	11.23
Quartz	165.62	12.07
Quicksilver	849.75	2.35
Cinnabar	531.25	3.76
Sulphur	127.12	15.74
Tin	459.00	4.35
Oxide	418.75	4.77
Zinc	437.50	4.57
Blende	253.12	7.90
Carbonate	273.12	7.32
Silicate	215.62	9.28
Andesite	165.62	12.07
Granite	162.62	12.30
Diabase	181.25	11.03
Diorite	171.87	11.63
Slates	165.62	12.07
Sandstones	162.50	12.30
Rhyolite	156.25	12.80

The specific gravity of any particular mineral has a considerable range, and a medium has been taken. The possible error is inconsequential for the purpose of these calculations.

For example, a representative gold ore may contain in the main 96% quartz, and 4% iron pyrite, and the weight of the ore may be deduced as follows:—

Quartz,	96% ×	12.0 7	11.58
Iron Pyrite,	4% ×	6.40	.25
			11.83 cubic feet per ton.

Most engineers, to compensate porosity, would allow twelve to thirteen cubic feet per ton.

CLASSIFICATION OF ORE IN SIGHT.

The risk in estimates of the average value of standing ore is dependent largely upon how far values disclosed by sampling are assumed to penetrate beyond the tested face, and this depends upon the geological character of the deposit. From theoretical grounds and experience, it is known that such values will have some extension, and the assumption of any given distance is a calculation of risk. The multiplication of development openings results in an increase of sampling points available and lessens the hazards. The frequency of such openings varies in different portions of every mine, and thus there are inequalities of risk. It is therefore customary in giving estimates of standing ore to classify the ore according to the degree of risk assumed, either by stating the number of sides exposed or by other phrases. Much discussion and ink have been devoted to trying to define what risk may be taken in such matters, that is in reality how far values may be assumed to penetrate into the unbroken ore. Still more has been consumed in attempts to coin terms and make classifications which will indicate what ratio of hazard has been taken in stating quantities and values.

The old terms "ore in sight" and "profit in sight" have been of late years subject to much malediction on the part of engineers because these expressions have been so badly abused by the charlatans of mining in attempts to cover the flights of their imaginations. A large part of Volume X of the "Institution of Mining and Metallurgy" has been devoted to heaping infamy on these terms, yet not only have they preserved their places in professional nomenclature, but nothing has been found to supersede them.

Some general term is required in daily practice to cover the whole field of visible ore, and if the phrase "ore in sight" be defined, it will be easier to teach the laymen its proper use than to abolish it. In fact, the substitutes are becoming abused as much as the originals ever were. All convincing expressions will be misused by somebody.

The legitimate direction of reform has been to divide the general term of "ore in sight" into classes, and give them names which will indicate the variable amount of risk of continuity in different parts of the mine. As the frequency of sample points, and consequently the risk of continuity, will depend upon the detail with which the mine is cut into blocks by the development openings, and upon the number of sides of such blocks which are accessible, most classifications of the degree of risk of continuity have been defined in terms of the number of sides exposed in the blocks. Many phrases have been coined to express such classifications; those most currently used are the following:—

Positive Ore Ore Developed	Ore exposed on four sides in blocks of a size variously prescribed.
Ore Blocked Out	Ore exposed on three sides within reasonable distance of each other.
Probable Ore Ore Developing	Ore exposed on two sides.
Possible Ore Ore Expectant	The whole or a part of the ore below the lowest level or beyond the range of vision.

No two of these parallel expressions mean quite the same thing; each more or less overlies into another class, and in fact none of them is based upon a logical footing for such a classification. For example, values can be assumed to penetrate some distance from every sampled face, even if it be only ten feet, so that ore exposed on one side will show some "positive" or "developed" ore which, on the lines laid down above, might be "probable" or even "possible" ore. Likewise, ore may be "fully developed" or "blocked out" so far as it is necessary for stoping purposes with modern wide intervals between levels, and still be in blocks too large to warrant an assumption of continuity of values to their centers (Fig. 1). As to the third class of "possible" ore, it conveys an impression of tangibility to a nebulous hazard, and should never be used in connection with positive tonnages. This part of the mine's value comes under extension of the deposit a long distance beyond openings, which is a speculation and cannot be defined in absolute tons without exhaustive explanation of the risks attached, in which case any phrase intended to shorten description is likely to be misleading.

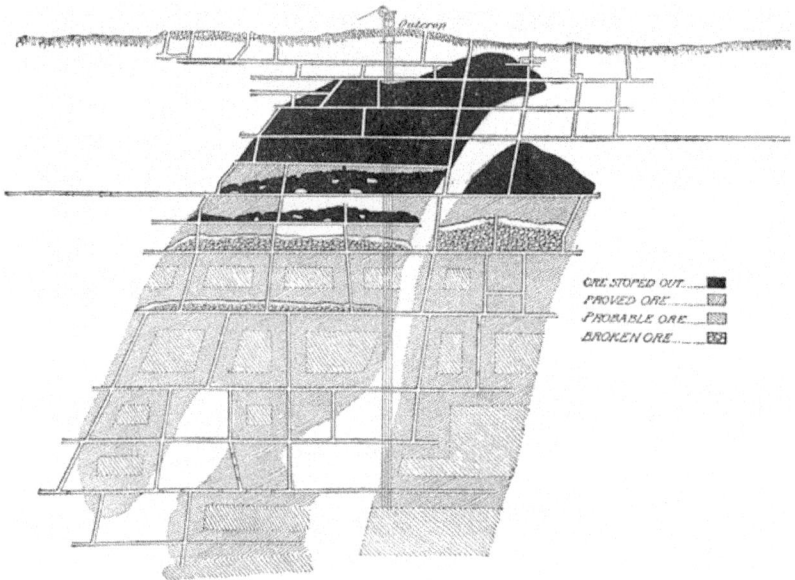

FIG. 1.—Longitudinal section of a mine, showing classification of the exposed ore. Scale, 400 feet = 1 inch.

Therefore empirical expressions in terms of development openings cannot be made to cover a geologic factor such as the distribution of metals through a rock mass. The only logical basis of ore classification for estimation purposes is one which is founded on the chances of the values penetrating from the surface of the exposures for each particular mine. Ore that may be calculated upon to a certainty is that which, taking into consideration the character of the deposit, can be said to be so sufficiently surrounded by sampled faces that the distance into the mass to which values are assumed to extend is reduced to a minimum risk. Ore so far removed from the sampled face as to leave some doubt, yet affording great reason for expectation of continuity, is "probable" ore. The third class of ore mentioned, which is that depending upon extension of the deposit and in which, as said above, there is great risk, should be treated separately as the speculative value of the mine. Some expressions are desirable for these classifications, and the writer's own preference is for the following, with a definition based upon the controlling factor itself.

They are:—

Proved Ore	Ore where there is practically no risk of failure of continuity.
Probable Ore	Ore where there is some risk, yet warrantable justification for assumption of continuity.
Prospective Ore	Ore which cannot be included in the above classes, nor definitely known or stated in any terms of tonnage.

What extent of openings, and therefore of sample faces, is required for the ore to be called "proved" varies naturally with the type of deposit,—in fact with each mine. In a general way, a fair rule in gold quartz veins below influence of secondary alteration is that no point in the block shall be over fifty feet from the points sampled. In limestone or andesite replacements, as by gold or lead or copper, the radius must be less. In defined lead and copper lodes, or in large lenticular bodies such as the Tennessee copper mines, the radius may often be considerably greater,—say one hundred feet. In gold deposits of such extraordinary regularity of values as the Witwatersrand bankets, it can well be two hundred or two hundred and fifty feet.

"Probable ore" should be ore which entails continuity of values through a greater distance than the above, and such distance must depend upon the collateral evidence from the character of the deposit, the position of openings, etc.

Ore beyond the range of the "probable" zone is dependent upon the extension of the deposit beyond the realm of development and will be discussed separately.

Although the expression "ore in sight" may be deprecated, owing to its abuse, some general term to cover both "positive" and "probable" ore is desirable; and where a general term is required, it is the intention herein to hold to the phrase "ore in sight" under the limitations specified.

CHAPTER III.

MINE VALUATION (*Continued*).

PROSPECTIVE VALUE.[1] EXTENSION IN DEPTH;
ORIGIN AND STRUCTURAL CHARACTER OF THE
DEPOSIT; SECONDARY ENRICHMENT;
DEVELOPMENT IN NEIGHBORING MINES; DEPTH OF
EXHAUSTION.

It is a knotty problem to value the extension of a deposit
beyond a short distance from the last opening. A short distance
beyond it is "proved ore," and for a further short distance is
"probable ore." Mines are very seldom priced at a sum so moderate as that represented by the profit to be won from the ore in
sight, and what value should be assigned to this unknown portion of the deposit admits of no certainty. No engineer can
approach the prospective value of a mine with optimism, yet the
mining industry would be non-existent to-day were it ap-
proached with pessimism. Any value assessed must be a matter
of judgment, and this judgment based on geological evidence.
Geology is not a mathematical science, and to attach a money
equivalent to forecasts based on such evidence is the most difficult task set for the mining engineer. It is here that his view of
geology must differ from that of his financially more irresponsi-
ble brother in the science. The geologist, contributing to human
knowledge in general, finds his most valuable field in the examination of mines largely exhausted. The engineer's most
valuable

[1] The term "extension in depth" is preferred by many to the phrase "prospective
value." The former is not entirely satisfactory, as it has a more specific than general
application. It is, however, a current miner's phrase, and is more expressive. In this
discussion "extension in depth" is used synonymously, and it may be taken to include
not alone the downward prolongation of the ore below workings, but also the occasional cases of lateral extension beyond the range of development work. The
commonest instance is continuance below the bottom level. In any event, to the majority of cases of different extension the same reasoning applies.

work arises from his ability to anticipate in the youth of the mine the symptoms of its old age. The work of our geologic friends is, however, the very foundation on which we lay our forecasts.

Geologists have, as the result of long observation, propounded for us certain hypotheses which, while still hypotheses, have proved to account so widely for our underground experience that no engineer can afford to lose sight of them. Although there is a lack of safety in fixed theories as to ore deposition, and although such conclusions cannot be translated into feet and metal value, they are nevertheless useful weights on the scale where probabilities are to be weighed.

A method in vogue with many engineers is, where the bottom level is good, to assume the value of the extension in depth as a sum proportioned to the profit in sight, and thus evade the use of geological evidence. The addition of various percentages to the profit in sight has been used by engineers, and proposed in technical publications, as varying from 25 to 50%. That is, they roughly assess the extension in depth to be worth one-fifth to one-third of the whole value of an equipped mine. While experience may have sometimes demonstrated this to be a practical method, it certainly has little foundation in either science or logic, and the writer's experience is that such estimates are untrue in practice. The quantity of ore which may be in sight is largely the result of managerial policy. A small mill on a large mine, under rapid development, will result in extensive ore-reserves, while a large mill eating away rapidly on the same mine under the same scale of development would leave small reserves. On the above scheme of valuation the extension in depth would be worth very different sums, even when the deepest level might be at the same horizon in both cases. Moreover, no mine starts at the surface with a large amount of ore in sight. Yet as a general rule this is the period when its extension is most valuable, for when the deposit is exhausted to 2000 feet, it is not likely to have such extension in depth as when opened one hundred feet, no matter what the ore-reserves may be. Further, such bases of valuation fail to take into account the widely varying geologic character of different mines, and they disregard any collateral evidence either of continuity from neighboring development, or from experience in the district. Logically, the prospective value can be simply a factor of how *far* the ore in the individual mine may be expected to extend, and not a factor of the remnant of ore that may still be unworked above the lowest level.

An estimation of the chances of this extension should be based solely on the local factors which bear on such extension, and these are almost wholly dependent upon the character of the deposit. These various geological factors from a mining engineer's point of view are:—

1. The origin and structural character of the ore-deposit.
2. The position of openings in relation to secondary alteration.
3. The size of the deposit.
4. The depth to which the mine has already been exhausted.
5. The general experience of the district for continuity and the development of adjoining mines.

The Origin and Structural Character of the Deposit.—In a general way, the ore-deposits of the order under discussion originate primarily through the deposition of metals from gases or solutions circulating along avenues in the earth's crust.[1] The original source of metals is a matter of great disagreement, and does not much concern the miner. To him, however, the origin and character of the avenue of circulation, the enclosing rock, the influence of the rocks on the solution, and of the solutions on the rocks, have a great bearing on the probable continuity of the volume and value of the ore.

All ore-deposits vary in value and, in the miner's view, only those portions above the pay limit are ore-bodies, or ore-shoots. The localization of values into such pay areas in an ore-deposit are apparently influenced by:

1. The distribution of the open spaces created by structural movement, fissuring, or folding as at Bendigo.

---

[1] The class of magmatic segregations is omitted, as not being of sufficiently frequent occurrence in payable mines to warrant troubling with it here.

2. The intersection of other fractures which, by mingling of solutions from different sources, provided precipitating conditions, as shown by enrichments at cross-veins.
3. The influence of the enclosing rocks by:—
    a. Their solubility, and therefore susceptibility to replacement.
    b. Their influence as a precipitating agent on solutions.
    c. Their influence as a source of metal itself.
    d. Their texture, in its influence on the character of the fracture. In homogeneous rocks the tendency is to open clean-cut fissures; in friable rocks, zones of brecciation; in slates or schistose rocks, linked lenticular open spaces;—these influences exhibiting themselves in miner's terms respectively in "well-defined fissure veins," "lodes," and "lenses."
    e. The physical character of the rock mass and the dynamic forces brought to bear upon it. This is a difficult study into the physics of stress in cases of fracturing, but its local application has not been without results of an important order.
4. Secondary alteration near the surface, more fully discussed later.

It is evident enough that the whole structure of the deposit is a necessary study, and even a digest of the subject is not to be compressed into a few paragraphs.

From the point of view of continuity of values, ore-deposits may be roughly divided into three classes. They are:—

1. Deposits of the infiltration type in porous beds, such as Lake Superior copper conglomerates and African gold bankets.
2. Deposits of the fissure vein type, such as California quartz veins.
3. Replacement or impregnation deposits on the lines of fissuring or otherwise.

In a general way, the uniformity of conditions of deposition in the first class has resulted in the most satisfactory continuity of ore and of its metal contents. In the second, depending much upon the profundity of the earth movements involved, there is laterally and vertically a reasonable basis for expectation of continuity but through much less distance than in the first class.

The third class of deposits exhibits widely different phenomena as to continuity and no generalization is of any value. In gold deposits of this type in West Australia, Colorado, and Nevada, continuity far beyond a sampled face must be received with the greatest skepticism. Much the same may be said of most copper replacements in limestone. On the other hand the most phenomenal regularity of values have been shown in certain Utah and Arizona copper mines, the result of secondary infiltration in porphyritic gangues. The Mississippi Valley lead and zinc deposits, while irregular in detail, show remarkable continuity by way of reoccurrence over wide areas. The estimation of the prospective value of mines where continuity of production is dependent on reoccurrence of ore-bodies somewhat proportional to the area, such as these Mississippi deposits or to some extent as in Cobalt silver veins, is an interesting study, but one that offers little field for generalization.

The Position of the Openings in Relation to Secondary Alteration.—The profound alteration of the upper section of ore-deposits by oxidation due to the action of descending surface waters, and their associated chemical agencies, has been generally recognized for a great many years. Only recently, however, has it been appreciated that this secondary alteration extends into the sulphide zone as well. The bearing of the secondary alteration, both in the oxidized and upper sulphide zones, is of the most sweeping economic character. In considering extension of values in depth, it demands the most rigorous investigation. Not only does the metallurgical character of the ores change with oxidation, but the complex reactions due to descending surface waters cause leaching and a migration of metals from one horizon to another lower down, and also in many cases a redistribution of their sequence in the upper zones of the deposit.

The effect of these agencies has been so great in many cases as to entirely alter the character of the mine and extension in depth has necessitated a complete re-equipment. For instance, the Mt. Morgan gold mine, Queensland, has now become a copper mine; the copper mines at Butte were formerly silver mines; Leadville has become largely a zinc producer instead of lead.

From this alteration aspect ore-deposits may be considered to have four horizons:—

1. The zone near the outcrop, where the dominating feature is oxidation and leaching of the soluble minerals.
2. A lower horizon, still in the zone of oxidation, where the predominant feature is the deposition of metals as native, oxides, and carbonates.
3. The upper horizon of the sulphide zone, where the special feature is the enrichment due to secondary deposition as sulphides.
4. The region below these zones of secondary alteration, where the deposit is in its primary state.

These zones are seldom sharply defined, nor are they always all in evidence. How far they are in evidence will depend, among other things, upon the amount and rapidity of erosion, the structure and mineralogical character of the deposit, and upon the enclosing rock.

If erosion is extremely rapid, as in cold, wet climates, and rough topography, or as in the case of glaciation of the Lake copper deposits, denudation follows close on the heels of alteration, and the surface is so rapidly removed that we may have the primary ore practically at the surface. Flat, arid regions present the other extreme, for denudation is much slower, and conditions are most perfect for deep penetration of oxidizing agencies, and the consequent alteration and concentration of the metals.

The migration of metals from the top of the oxidized zone

leaves but a barren cap for erosion. The consequent effect of denudation that lags behind alteration is to raise slowly the concentrated metals toward the surface, and thus subject them to renewed attack and repeated migration. In this manner we can account for the enormous concentration of values in the lower oxidized and upper sulphide zones overlying very lean sulphides in depth.

Some minerals are more freely soluble and more readily precipitated than others. From this cause there is in complex metal deposits a rearrangement of horizontal sequence, in addition to enrichment at certain horizons and impoverishment at others. The whole subject is one of too great complexity for adequate consideration in this discussion. No engineer is properly equipped to give judgment on extension in depth without a thorough grasp of the great principles laid down by Van Hise, Emmons, Lindgren, Weed, and others. We may, however, briefly examine some of the theoretical effects of such alteration.

Zinc, iron, and lead sulphides are a common primary combination. These metals are rendered soluble from their usual primary forms by oxidizing agencies, in the order given. They reprecipitate as sulphides in the reverse sequence. The result is the leaching of zinc and iron readily in the oxidized zone, thus differentially enriching the lead which lags behind, and a further extension of the lead horizon is provided by the early precipitation of such lead as does migrate. Therefore, the lead often predominates in the second and the upper portion of the third zone, with the zinc and iron below. Although the action of all surface waters is toward oxidation and carbonation of these metals, the carbonate development of oxidized zones is more marked when the enclosing rocks are calcareous.

In copper-iron deposits, the comparatively easy decomposition and solubility and precipitation of the copper and some iron salts generally result in more extensive impoverishment of these metals near the surface, and more predominant enrichment at a lower horizon than is the case with any other metals. The barren "iron hat" at the first zone, the carbonates and oxides at the second, the enrichment with secondary copper sulphides at the top of the third, and the occurrence of secondary copper-iron sulphides below, are often most clearly defined. In the easy recognition of the secondary copper sulphides, chalcocite, bornite, etc., the engineer finds a finger-post on the road to extension in depth; and the directions upon this post are not to be disregarded. The number of copper deposits enriched from unpayability in the first zone to a profitable character in the next two, and unpayability again in the fourth, is legion.

Silver occurs most abundantly in combination with either lead, copper, iron, or gold. As it resists oxidation and solution more strenuously than copper and iron, its tendency when in combination with them is to lag behind in migration. There is thus a differential enrichment of silver in the upper two zones, due to the reduction in specific gravity of the ore by the removal of associated metals. Silver does migrate somewhat, however, and as it precipitates more readily than copper, lead, zinc, or iron, its tendency when in combination with them is towards enrichment above the horizons of enrichment of these metals. When it is in combination with lead and zinc, its very ready precipitation from solution by the galena leaves it in combination more predominantly with the lead. The secondary enrichment of silver deposits at the top of the sulphide zone is sometimes a most pronounced feature, and it seems to be the explanation of the origin of many "bonanzas."

In gold deposits, the greater resistance to solubility of this metal than most of the others, renders the phenomena of migration to depth less marked. Further than this, migration is often interfered with by the more impervious quartz matrix of many gold deposits. Where gold is associated with large quantities of base metals, however, the leaching of the latter in the oxidized zone leaves the ore differentially richer, and as gold is also slightly soluble, in such cases the migration of the base metals does carry some of the gold. In the instance especially of impregnation or replacement deposits, where the matrix is easily permeable, the upper sulphide zone is distinctly richer than lower down, and this enrichment is accompanied by a considerable increase in sulphides and tellu-rides. The predominant characteristic of alteration in gold deposits is, however, enrichment in the oxidized zone with the maximum values near the surface. The reasons for this appear to be that gold in its resistance to oxidation and wholesale migra-tion gives opportunities to a sort of combined mechanical and chemical enrichment.

In dry climates, especially, the gentleness of erosion allows of more thorough decomposition of the outcroppings, and a me-chanical separation of the gold from the detritus. It remains on or near the deposit, ready to be carried below, mechanically or oth-erwise. In wet climates this is less pronounced, for erosion bears away the croppings before such an extensive decomposition and freeing of the gold particles. The West Australian gold fields present an especially prominent example of this type of superfi-cial enrichment. During the last fifteen years nearly eight hundred companies have been formed for working mines in this region. Although from four hundred of these high-grade ore has been produced, some thirty-three only have ever paid dividends. The great majority have been unpayable below oxidation,—a distance of one or two hundred feet. The writer's unvarying ex-perience with gold is that it is richer in the oxidized zone than at any point below. While cases do occur of gold deposits richer in the upper sulphide zone than below, even the upper sulphides are usually poorer than the oxidized region. In quartz veins pre-eminently, evidence of enrichment in the third zone is likely to be practically absent.

Tin ores present an anomaly among the base metals under dis-cussion, in that the primary form of this metal in most workable deposits is an oxide. Tin in this form is most difficult of solution from ground agencies, as witness the great alluvial deposits, of-ten of considerable geologic age. In consequence the phenomena of migration and enrichment are almost wholly absent, except such as are due to mechanical penetration of tin from surface decomposition of the matrix akin to that described in gold depo-sits.

In general, three or four essential facts from secondary alteration must be kept in view when prognosticating extensions.

Oxidation usually alters treatment problems, and oxidized ore of the same grade as sulphides can often be treated more cheaply. This is not universal. Low-grade ores of lead, copper, and zinc may be treatable by concentration when in the form of sulphides, and may be valueless when oxidized, even though of the same grade.

Copper ores generally show violent enrichment at the base of the oxidized, and at the top of the sulphide zone.

Lead-zinc ores show lead enrichment and zinc impoverishment in the oxidized zone but have usually less pronounced enrichment below water level than copper. The rearrangement of the metals by the deeper migration of the zinc, also renders them metallurgically of less value with depth.

Silver deposits are often differentially enriched in the oxidized zone, and at times tend to concentrate in the upper sulphide zone.

Gold deposits usually decrease in value from the surface through the whole of the three alteration zones.

Size of Deposits.—The proverb of a relation between extension in depth and size of ore-bodies expresses one of the oldest of miners' beliefs. It has some basis in experience, especially in fissure veins, but has little foundation in theory and is applicable over but limited areas and under limited conditions.

From a structural view, the depth of fissuring is likely to be more or less in proportion to its length and breadth and therefore the volume of vein filling with depth is likely to be proportional to length and width of the fissure. As to the distribution of values, if we eliminate the influence of changing wall rocks, or other precipitating agencies which often cause the values to arrange themselves in "floors," and of secondary alteration, there may be some reason to assume distribution of values of an extent equal vertically to that displayed horizontally. There is, as said, more reason in experience for this assumption than in theory. A study of the shape of a great many ore-shoots in mines of fissure type indicates that when the ore-shoots or ore-bodies are approaching vertical exhaustion they do not end abruptly, but gradually shorten and decrease in value, their bottom boundaries being more often wedge-shaped than even lenticular. If this could be taken as the usual occurrence, it would be possible (eliminating the evident exceptions mentioned above) to state roughly that the minimum extension of an ore-body or ore-shoot in depth below any given horizon would be a distance represented by a radius equal to one-half its length. By length is not meant necessarily the length of a horizontal section, but of one at right angles to the downward axis.

On these grounds, which have been re-enforced by much experience among miners, the probabilities of extension are somewhat in proportion to the length and width of each ore-body. For instance, in the A mine, with an ore-shoot 1000 feet long and 10 feet wide, on its bottom level, the minimum extension under this hypothesis would be a wedge-shaped ore-body with its deepest point 500 feet below the lowest level, or a minimum of say 200,000 tons. Similarly, the B mine with five ore-bodies, each 300 hundred feet long and 10 feet wide, exposed on its lowest level, would have a minimum of five wedges 100 feet deep at their deepest points, or say 50,000 tons. This is not proposed as a formula giving the total amount of extension in depth, but as a sort of yardstick which has experience behind it. This experience applies in a much less degree to deposits originating from impregnation along lines of fissuring and not at all to replacements.

Development in Neighboring Mines.—Mines of a district are usually found under the same geological conditions, and show somewhat the same habits as to extension in depth or laterally, and especially similar conduct of ore-bodies and ore-shoots.

As a practical criterion, one of the most intimate guides is the actual development in adjoining mines. For instance, in Kalgoorlie, the Great Boulder mine is (March, 1908) working the extension of Ivanhoe lodes at points 500 feet below the lowest level in the Ivanhoe; likewise, the Block 10 lead mine at Broken Hill is working the Central ore-body on the Central boundary some 350 feet below the Central workings. Such facts as these must have a bearing on assessing the downward extension.

Depth of Exhaustion.—All mines become completely exhausted at some point in depth. Therefore the actual distance to which ore can be expected to extend below the lowest level grows less with every deeper working horizon. The really superficial character of ore-deposits, even outside of the region of secondary enrichment is becoming every year better recognized. The prospector's idea that "she gets richer deeper down," may have some basis near the surface in some metals, but it is not an idea which prevails in the minds of engineers who have to work in depth. The writer, with some others, prepared a list of several hundred dividend-paying metal mines of all sorts, extending over North and South America, Australasia, England, and Africa. Notes were made as far as possible of the depths at which values gave out, and also at which dividends ceased. Although by no means a complete census, the list indicated that not 6% of mines (outside banket) that have yielded profits, ever made them from ore won below 2000 feet. Of mines that paid dividends, 80% did not show profitable value below 1500 feet, and a sad majority died above 500. Failures at short depths may be blamed upon secondary enrichment, but the majority that reached below this influence also gave out. The geological reason for such general unseemly conduct is not so evident.

Conclusion.—As a practical problem, the assessment of prospective value is usually a case of "cut and try." The portion of the capital to be invested, which depends upon extension, will require so many tons of ore of the same value as that indicated by the standing ore, in order to justify the price.

To produce this tonnage at the continued average size of the ore-bodies will require their extension in depth so many feet—or the discovery of new ore-bodies of a certain size. The five geological weights mentioned above may then be put into the scale and a basis of judgment reached.

CHAPTER IV.

MINE VALUATION (*Continued*).

RECOVERABLE PERCENTAGE OF THE GROSS ASSAY
VALUE; PRICE OF METALS; COST OF PRODUCTION.

The method of treatment for the ore must be known before a
mine can be valued, because a knowledge of the recoverable
percentage is as important as that of the gross value of the ore
itself. The recoverable percentage is usually a factor of working
costs. Practically every ore can be treated and all the metal contents recovered, but the real problem is to know the method and
percentage of recovery which will yield the most remunerative
result, if any. This limit to profitable recovery regulates the
amount of metal which should be lost, and the amount of metal
which consequently must be deducted from the gross value before the real net value of the ore can be calculated. Here, as
everywhere else in mining, a compromise has to be made with
nature, and we take what we can get—profitably. For instance, a
copper ore may be smelted and a 99% recovery obtained. Under
certain conditions this might be done at a loss, while the same
ore might be concentrated before smelting and yield a profit with
a 70% recovery. An additional 20% might be obtained by roast-
ing and leaching the residues from concentration, but this would
probably result in an expenditure far greater than the value of the
20% recovered. If the ore is not already under treatment on the
mine, or exactly similar ore is not under treatment elsewhere,
with known results, the method must be determined experimen-
tally, either by the examining engineer or by a special
metallurgist.

Where partially treated products, such as concentrates, are to
be sold, not only will there be further losses, but

deductions will be made by the smelter for deleterious metals and other charges. All of these factors must be found out,—and a few sample smelting returns from a similar ore are useful.

To cover the whole field of metallurgy and discuss what might apply, and how it might apply, under a hundred supposititious conditions would be too great a digression from the subject in hand. It is enough to call attention here to the fact that the residues from every treatment carry some metal, and that this loss has to be deducted from the gross value of the ore in any calculations of net values.

PRICE OF METALS.

Unfortunately for the mining engineer, not only has he to weigh the amount of risk inherent in calculations involved in the mine itself, but also that due to fluctuations in the value of metals. If the ore is shipped to custom works, he has to contemplate also variations in freights and smelting charges. Gold from the mine valuer's point of view has no fluctuations. It alone among the earth's products gives no concern as to the market price. The price to be taken for all other metals has to be decided before the mine can be valued. This introduces a further speculation and, as in all calculations of probabilities, amounts to an estimate of the amount of risk. In a free market the law of supply and demand governs the value of metals as it does that of all other commodities. So far, except for tariff walls and smelting rings, there is a free market in the metals under discussion.

The demand for metals varies with the unequal fluctuations of the industrial tides. The sea of commercial activity is subject to heavy storms, and the mine valuer is compelled to serve as weather prophet on this ocean of trouble. High prices, which are the result of industrial booms, bring about overproduction, and the collapse of these begets a shrinkage of demand, wherein consequently the tide of price turns back. In mining for metals each pound is produced actually at a different cost. In case of an oversupply of base metals the price will fall until it has reached a point where a portion of the production is no longer profitable, and the equilibrium is established through decline in output. However, in the backward swing, due to lingering overproduction, prices usually fall lower than the cost of producing even a much-diminished supply. There is at this point what we may call the "basic" price, that at which production is insufficient and the price rises again. The basic price which is due to this undue backward swing is no more the real price of the metal to be contemplated over so long a term of years than is the highest price. At how much above the basic price of depressed times the product can be safely expected to find a market is the real question. Few mines can be bought or valued at this basic price. An indication of what this is can be gained from a study of fluctuations over a long term of years.

It is common to hear the average price over an extended period considered the "normal" price, but this basis for value is one which must be used with discretion, for it is not the whole question when mining. The "normal" price is the average price over a long term. The lives of mines, and especially ore in sight, may not necessarily enjoy the period of this "normal" price. The engineer must balance his judgments by the immediate outlook of the industrial weather. When lead was falling steadily in December, 1907, no engineer would accept the price of that date, although it was then below "normal"; his product might go to market even lower yet.

It is desirable to ascertain what the basic and normal prices are, for between them lies safety. Since 1884 there have been three cycles of commercial expansion and contraction. If the average prices are taken for these three cycles separately (1885-95), 1895-1902, 1902-08) it will be seen that there has been a steady advance in prices. For the succeeding cycles lead on the London Exchange,[1] the freest of the world's

[1] All London prices are based on the long ton of 2,240 lbs. Much confusion exists in the copper trade as to the classification of the metal. New York prices are quoted in electrolytic and "Lake"; London's in "Standard." "Standard" has now become practically an arbitrary term peculiar to London, for the great bulk of copper dealt in is "electrolytic" valued considerably over "Standard."]

markets was £12 12*s.* 4*d.*, £13 3*s.* 7*d.*, and £17 7*s.* 0*d.* respectively; zinc, £17 14*s.* 10*d.*, £19 3*s.*8*d.*, and £23 3*s.* 0*d.*; and standard copper,£48 16*s.* 0*d.*, £59 10*s.* 0*d.*, and £65 7*s.* 0*d.* It seems, therefore, that a higher standard of prices can be assumed as the basic and normal than would be indicated if the general average of, say, twenty years were taken. During this period, the world's gold output has nearly quadrupled, and, whether the quantitative theory of gold be accepted or not, it cannot be denied that there has been a steady increase in the price of commodities. In all base-metal mining it is well to remember that the production of these metals is liable to great stimulus at times from the discovery of new deposits or new processes of recovery from hitherto unprofitable ores. It is therefore for this reason hazardous in the extreme to prophesy what prices will be far in the future, even when the industrial weather is clear. But some basis must be arrived at, and from the available outlook it would seem that the following metal prices are justifiable for some time to come, provided the present tariff schedules are maintained in the United States:

	LEAD		SPELTER		COPPER		TIN		SILVER	
	London Ton	N.Y. Pound	Lon. Ton	N.Y. Pound	Lon. Ton	N.Y. Pound	Lon. Ton	N.Y. Pound	Lon. Per oz.	N.Y. Per oz.
Basic Price	£11.	$.035	£17	$.040	£52	$.115	£100	$.220	22*d.*	$.44
Normal Price	13.5	.043	21	.050	65	.140	130	.290	26	.52

In these figures the writer has not followed strict averages, but has taken the general outlook combined with the previous records. The likelihood of higher prices for lead is more encouraging than for any other metal, as no new deposits of importance have come forward for years, and the old mines are reaching considerable depths. Nor does the frenzied prospecting of the world's surface during the past ten years appear to forecast any very disturbing developments. The zinc future is not so bright, for metallurgy has done wonders in providing methods of saving the zinc formerly discarded from lead ores, and enormous supplies will come forward when required. The tin outlook is encouraging, for the supply from a mining point of view seems unlikely to more than keep pace with the world's needs. In copper the demand is growing prodigiously, but the supplies of copper ores and the number of copper mines that are ready to produce whenever normal prices recur was never so great as to-day. One very hopeful fact can be deduced for the comfort of the base metal mining industry as a whole. If the growth of demand continues through the next thirty years in the ratio of the past three decades, the annual demand for copper will be over 3,000,000 tons, of lead over 1,800,000 tons, of spelter 2,800,000 tons, of tin 250,000 tons. Where such stupendous amounts of these metals are to come from at the present range of prices, and even with reduced costs of production, is far beyond any apparent source of supply. The outlook for silver prices is in the long run not bright. As the major portion of the silver produced is a bye product from base metals, any increase in the latter will increase the silver production despite very much lower prices for the precious metal. In the meantime the gradual conversion of all nations to the gold standard seems a matter of certainty. Further, silver may yet be abandoned as a subsidiary coinage inasmuch as it has now but a token value in gold standard countries if denuded of sentiment.

COST OF PRODUCTION.

It is hardly necessary to argue the relative importance of the determination of the cost of production and the determination of the recoverable contents of the ore. Obviously, the aim of mine valuation is to know the profits to be won, and the profit is the value of the metal won, less the cost of production.

The cost of production embraces development, mining, treatment, management. Further than this, it is often contended that, as the capital expended in purchase and equipment must be redeemed within the life of the mine, this item should also be included in production costs. It is true that mills, smelters, shafts, and all the paraphernalia of a mine are of virtually negligible value when it is exhausted; and that all mines are exhausted sometime and every ton taken out contributes to that exhaustion; and that every ton of ore must bear its contribution to the return of the investment, as well as profit upon it. Therefore it may well be said that the redemption of the capital and its interest should be considered in costs per ton. The difficulty in dealing with the subject from the point of view of production cost arises from the fact that, except possibly in the case of banket gold and some conglomerate copper mines, the life of a metal mine is unknown beyond the time required to exhaust the ore reserves. The visible life at the time of purchase or equipment may be only three or four years, yet the average equipment has a longer life than this, and the anticipation for every mine is also for longer duration than the bare ore in sight. For clarity of conclusions in mine valuation the most advisable course is to determine the profit in sight irrespective of capital redemption in the first instance. The questions of capital redemption, purchase price, or equipment cost can then be weighed against the margin of profit. One phase of redemption will be further discussed under "Amortization of Capital" and "Ratio of Output to the Mine."

The cost of production depends upon many things, such as the cost of labor, supplies, the size of the ore-body, the treatment necessary, the volume of output, etc.; and to discuss them all would lead into a wilderness of supposititious cases. If the mine is a going concern, from which reliable data can be obtained, the problem is much simplified. If it is virgin, the experience of other mines in the same region is the next resource; where no such data can be had, the engineer must fall back upon the experience with mines still farther afield. Use is sometimes made of the "comparison ton" in calculating costs upon mines where data of actual experience are not available. As costs will depend in the main upon items mentioned above, if the known costs of a going mine elsewhere be taken as a basis, and subtractions and additions made for more unfavorable or favorable effect of the differences in the above items, a fairly close result can be approximated.

Mine examinations are very often inspired by the belief that extended operations or new metallurgical applications to the mine will expand the profits. In such cases the paramount questions are the reduction of costs by better plant, larger outputs, new processes, or alteration of metallurgical basis and better methods. If every item of previous expenditure be gone over and considered, together with the equipment, and method by which it was obtained, the possible savings can be fairly well deduced, and justification for any particular line of action determined. One view of this subject will be further discussed under "Ratio of Output to the Mine." The conditions which govern the working costs are on every mine so special to itself, that no amount of advice is very useful. Volumes of advice have been published on the subject, but in the main their burden is not to underestimate.

In considering the working costs of base-metal mines, much depends upon the opportunity for treatment in customs works, smelters, etc. Such treatment means a saving of a large portion of equipment cost, and therefore of the capital to be invested and subsequently recovered. The economics of home treatment must be weighed against the sum which would need to be set aside for redemption of the plant, and unless there is a very distinct advantage to be had by the former, no risks should be taken. More engineers go wrong by the erection of treatment works where other treatment facilities are available, than do so by continued shipping. There are many mines where the cost of equipment could never be returned, and which would be valueless unless the ore could be shipped. Another phase of foreign treatment arises from the necessity or advantage of a mixture of ores,—the opportunity of such mixtures often gives the public smelter an advantage in treatment with which treatment on the mine could never compete.

Fluctuation in the price of base metals is a factor so much to be taken into consideration, that it is desirable in estimating mine values to reduce the working costs to a basis of a "per unit" of finished metal. This method has the great advantage of indicating so simply the involved risks of changing prices that whoso runs may read. Where one metal predominates over the other to such an extent as to form the "backbone" of the value of the mine, the value of the subsidiary metals is often deducted from the cost of the principal metal, in order to indicate more plainly the varying value of the mine with the fluctuating prices of the predominant metal. For example, it is usual to state that the cost of copper production from a given ore will be so many cents per pound, or so many pounds sterling per ton. Knowing the total metal extractable from the ore in sight, the profits at given prices of metal can be readily deduced. The point at which such calculation departs from the "per-ton-of-ore" unto the per-unit-cost-of-metal basis, usually lies at the point in ore dressing where it is ready for the smelter. To take a simple case of a lead ore averaging 20%: this is to be first concentrated and the lead reduced to a concentrate averaging 70% and showing a recovery of 75% of the total metal content. The cost per ton of development, mining, concentration, management, is to this point say $4 per ton of original crude ore. The smelter buys the concentrate for 95% of the value of the metal, less the smelting charge of $15 per ton, or there is a working cost of a similar sum by home equipment. In this case 4.66 tons of ore are required to produce one ton of concentrates, and therefore each ton of concentrates costs $18.64. This amount, added to the smelting charge, gives a total of $33.64 for the creation of 70% of one ton of finished lead, or equal to 2.40 cents per pound which can be compared with the market price less 5%. If the ore were to contain 20 ounces of silver per ton, of which 15 ounces were recovered into the leady concentrates, and the smelter price for the silver were 50 cents per ounce, then the $7.50 thus recovered would be subtracted from $33.64, making the apparent cost of the lead 1.86 cents per pound.

CHAPTER V.

MINE VALUATION (*Continued*).

REDEMPTION OR AMORTIZATION OF CAPITAL AND INTEREST.

It is desirable to state in some detail the theory of amortization before consideration of its application in mine valuation.

As every mine has a limited life, the capital invested in it must be redeemed during the life of the mine. It is not sufficient that there be a bare profit over working costs. In this particular, mines differ wholly from many other types of investment, such as railways. In the latter, if proper appropriation is made for maintenance, the total income to the investor can be considered as interest or profit; but in mines, a portion of the annual income must be considered as a return of capital. Therefore, before the yield on a mine investment can be determined, a portion of the annual earnings must be set aside in such a manner that when the mine is exhausted the original investment will have been restored. If we consider the date due for the return of the capital as the time when the mine is exhausted, we may consider the annual instalments as payments before the due date, and they can be put out at compound interest until the time for restoration arrives. If they be invested in safe securities at the usual rate of about 4%, the addition of this amount of compound interest will assist in the repayment of the capital at the due date, so that the annual contributions to a sinking fund need not themselves aggregate the total capital to be restored, but may be smaller by the deficiency which will be made up by their interest earnings. Such a system of redemption of capital is called "Amortization."

Obviously it is not sufficient for the mine investor that his capital shall have been restored, but there is required an excess earning over and above the necessities of this annual funding of capital. What rate of excess return the mine must yield is a matter of the risks in the venture and the demands of the investor. Mining business is one where 7% above provision for capital return is an absolute minimum demanded by the risks inherent in mines, even where the profit in sight gives warranty to the return of capital. Where the profit in sight (which is the only real guarantee in mine investment) is below the price of the investment, the annual return should increase in proportion. There are thus two distinct directions in which interest must be computed,—first, the internal influence of interest in the amortization of the capital, and second, the percentage return upon the whole investment after providing for capital return.

There are many limitations to the introduction of such refinements as interest calculations in mine valuation. It is a subject not easy to discuss with finality, for not only is the term of years unknown, but, of more importance, there are many factors of a highly speculative order to be considered in valuing. It may be said that a certain life is known in any case from the profit in sight, and that in calculating this profit a deduction should be made from the gross profit for loss of interest on it pending recovery. This is true, but as mines are seldom dealt with on the basis of profit in sight alone, and as the purchase price includes usually some proportion for extension in depth, an unknown factor is introduced which outweighs the known quantities. Therefore the application of the culminative effect of interest accumulations is much dependent upon the sort of mine under consideration. In most cases of uncertain continuity in depth it introduces a mathematical refinement not warranted by the speculative elements. For instance, in a mine where the whole value is dependent upon extension of the deposit beyond openings, and where an expected return of at least 50% per annum is required to warrant the risk, such refinement would be absurd. On the other hand, in a Witwatersrand gold mine, in gold and tin gravels, or in massive copper mines such as Bingham and Lake Superior, where at least some sort of life can be approximated, it becomes a most vital element in valuation.

In general it may be said that the lower the total annual return expected upon the capital invested, the greater does the amount demanded for amortization become in proportion to this total income, and therefore the greater need of its introduction in calculations. Especially is this so where the cost of equipment is large proportionately to the annual return. Further, it may be said that such calculations are of decreasing use with increasing proportion of speculative elements in the price of the mine. The risk of extension in depth, of the price of metal, etc., may so outweigh the comparatively minor factors here introduced as to render them useless of attention.

In the practical conduct of mines or mining companies, sinking funds for amortization of capital are never established. In the vast majority of mines of the class under discussion, the ultimate duration of life is unknown, and therefore there is no basis upon which to formulate such a definite financial policy even were it desired. Were it possible to arrive at the annual sum to be set aside, the stockholders of the mining type would prefer to do their own reinvestment. The purpose of these calculations does not lie in the application of amortization to administrative finance. It is nevertheless one of the touchstones in the valuation of certain mines or mining investments. That is, by a sort of inversion such calculations can be made to serve as a means to expose the amount of risk,—to furnish a yardstick for measuring the amount of risk in the very speculations of extension in depth and price of metals which attach to a mine. Given the annual income being received, or expected, the problem can be formulated into the determination of how many years it must be continued in order to amortize the investment and pay a given rate of profit. A certain length of life is evident from the ore in sight, which may be called the life in sight. If the term of years required to redeem the capital and pay an interest upon it is greater than the life in sight, then this extended life must come from extension in depth, or ore from other direction, or increased price of metals. If we then take the volume and profit on the ore as disclosed we can calculate the number of feet the deposit must extend in depth, or additional tonnage that must be obtained of the same grade, or the different prices of metal that must be secured, in order to satisfy the demanded term of years. These demands in actual measure of ore or feet or higher price can then be weighed against the geological and industrial probabilities.

The following tables and examples may be of assistance in these calculations.

Table 1. To apply this table, the amount of annual income or dividend and the term of years it will last must be known or estimated factors. It is then possible to determine the *present* value of this annual income after providing for amortization and interest on the investment at various rates given, by multiplying the annual income by the factor set out.

A simple illustration would be that of a mine earning a profit of $200,000 annually, and having a total of 1,000,000 tons in sight, yielding a profit of $2 a ton, or a total profit in sight of $2,000,000, thus recoverable in ten years. On a basis of a 7% return on the investment and amortization of capital (Table I), the factor is 6.52 x $200,000 = $1,304,000 as the present value of the gross profits exposed. That is, this sum of $1,304,000, if paid for the mine, would be repaid out of the profit in sight, together with 7% interest if the annual payments into sinking fund earn 4%.

TABLE I.

Present Value of an Annual Dividend Over —Years at — % and Replacing Capital by Reinvestment of an Annual Sum at 4%.

Years	5%	6%	7%	8%	9%	10%
1	.95	.94	.93	.92	.92	.91
2	1.85	1.82	1.78	1.75	1.72	1.69
3	2.70	2.63	2.56	2.50	2.44	2.38
4	3.50	3.38	3.27	3.17	3.07	2.98
5	4.26	4.09	3.93	3.78	3.64	3.51
6	4.98	4.74	4.53	4.33	4.15	3.99
7	5.66	5.36	5.09	4.84	4.62	4.41
8	6.31	5.93	5.60	5.30	5.04	4.79
9	6.92	6.47	6.08	5.73	5.42	5.14
10	7.50	6.98	6.52	6.12	5.77	5.45
11	8.05	7.45	6.94	6.49	6.09	5.74
12	8.58	7.90	7.32	6.82	6.39	6.00
13	9.08	8.32	7.68	7.13	6.66	6.24
14	9.55	8.72	8.02	7.42	6.91	6.46
15	10.00	9.09	8.34	7.79	7.14	6.67
16	10.43	9.45	8.63	7.95	7.36	6.86
17	10.85	9.78	8.91	8.18	7.56	7.03
18	11.24	10.10	9.17	8.40	7.75	7.19
19	11.61	10.40	9.42	8.61	7.93	7.34
20	11.96	10.68	9.65	8.80	8.09	7.49
21	12.30	10.95	9.87	8.99	8.24	7.62
22	12.62	11.21	10.08	9.16	8.39	7.74
23	12.93	11.45	10.28	9.32	8.52	7.85
24	13.23	11.68	10.46	9.47	8.65	7.96
25	13.51	11.90	10.64	9.61	8.77	8.06
26	13.78	12.11	10.80	9.75	8.88	8.16
27	14.04	12.31	10.96	9.88	8.99	8.25
28	14.28	12.50	11.11	10.00	9.09	8.33
29	14.52	12.68	11.25	10.11	9.18	8.41
30	14.74	12.85	11.38	10.22	9.27	8.49
31	14.96	13.01	11.51	10.32	9.36	8.56
32	15.16	13.17	11.63	10.42	9.44	8.62
33	15.36	13.31	11.75	10.51	9.51	8.69
34	15.55	13.46	11.86	10.60	9.59	8.75
35	15.73	13.59	11.96	10.67	9.65	8.80
36	15.90	13.72	12.06	10.76	9.72	8.86
37	16.07	13.84	12.16	10.84	9.78	8.91
38	16.22	13.96	12.25	10.91	9.84	8.96
39	16.38	14.07	12.34	10.98	9.89	9.00
40	16.52	14.18	12.42	11.05	9.95	9.05

Condensed from Inwood's Tables.

Table II is practically a compound discount table. That is, by it can be determined the present value of a fixed sum payable at the end of a given term of years, interest being discounted at various given rates. Its use may be illustrated by continuing the example preceding.

TABLE II.

PRESENT VALUE OF $1, OR £1, PAYABLE IN —YEARS, INTEREST TAKEN AT —%.

YEARS	4%	5%	6%	7%
1	.961	.952	.943	.934
2	.924	.907	.890	.873
3	.889	.864	.840	.816
4	.854	.823	.792	.763
5	.821	.783	.747	.713
6	.790	.746	.705	.666
7	.760	.711	.665	.623
8	.731	.677	.627	.582
9	.702	.645	.592	.544
10	.675	.614	.558	.508
11	.649	.585	.527	.475
12	.625	.557	.497	.444
13	.600	.530	.469	.415
14	.577	.505	.442	.388
15	.555	.481	.417	.362
16	.534	.458	.394	.339
17	.513	.436	.371	.316
18	.494	.415	.350	.296
19	.475	.396	.330	.276
20	.456	.377	.311	.258
21	.439	.359	.294	.241
22	.422	.342	.277	.266
23	.406	.325	.262	.211
24	.390	.310	.247	.197
25	.375	.295	.233	.184
26	.361	.281	.220	.172
27	.347	.268	.207	.161
28	.333	.255	.196	.150
29	.321	.243	.184	.140
30	.308	.231	.174	.131
31	.296	.220	.164	.123
32	.285	.210	.155	.115
33	.274	.200	.146	.107
34	.263	.190	.138	.100
35	.253	.181	.130	.094
36	.244	.172	.123	.087
37	.234	.164	.116	.082
38	.225	.156	.109	.076
39	.216	.149	.103	.071
40	.208	.142	.097	.067

Condensed from Inwood's Tables.

If such a mine is not equipped, and it is assumed that $200,000 are required to equip the mine, and that two years are required for this equipment, the value of the ore in sight is still less, because of the further loss of interest in delay and the cost of equipment. In this case the present value of $1,304,000 in two years, interest at 7%, the factor is .87 X 1,304,000 = $1,134,480. From this comes off the cost of equipment, or $200,000, leaving $934,480 as the present value of the profit in sight. A further refinement could be added by calculating the interest chargeable against the $200,000 equipment cost up to the time of production.

TABLE III.

Annual Rate of Dividend.	Number of years of life required to yield—% interest, and in addition to furnish annual instalments which, if reinvested at 4% will return the original investment at the end of the period.					
%	5%	6%	7%	8%	9%	10%
6	41.0					
7	28.0	41.0				
8	21.6	28.0	41.0			
9	17.7	21.6	28.0	41.0		
10	15.0	17.7	21.6	28.0	41.0	
11	13.0	15.0	17.7	21.6	28.0	41.0
12	11.5	13.0	15.0	17.7	21.6	28.0
13	10.3	11.5	13.0	15.0	17.7	21.6
14	9.4	10.3	11.5	13.0	15.0	17.7
15	8.6	9.4	10.3	11.5	13.0	15.0
16	7.9	8.6	9.4	10.3	11.5	13.0
17	7.3	7.9	8.6	9.4	10.3	11.5
18	6.8	7.3	7.9	8.6	9.4	10.3
19	6.4	6.8	7.3	7.9	8.6	9.4
20	6.0	6.4	6.8	7.3	7.9	8.6
21	5.7	6.0	6.4	6.8	7.3	7.9
22	5.4	5.7	6.0	6.4	6.8	7.3
23	5.1	5.4	5.7	6.0	6.4	6.8
24	4.9	5.1	5.4	5.7	6.0	6.4
25	4.7	4.9	5.1	5.4	5.7	6.0
26	4.5	4.7	4.9	5.1	5.4	5.7
27	4.3	4.5	4.7	4.9	5.1	5.4
28	4.1	4.3	4.5	4.7	4.9	5.1
29	3.9	4.1	4.3	4.5	4.7	4.9
30	3.8	3.9	4.1	4.3	4.5	4.7

Table III. This table is calculated by inversion of the factors in Table I, and is the most useful of all such tables, as it is a direct calculation of the number of years that a given rate of income on the investment must continue in order to amortize the capital (the annual sinking fund being placed at compound interest at 4%) and to repay various rates of interest on the investment. The application of this method in testing the value of dividend-paying shares is very helpful, especially in weighing the risks involved in the portion of the purchase or investment unsecured by the profit in sight. Given the annual percentage income on the investment from the dividends of the mine (or on a non-producing mine assuming a given rate of production and profit from the factors exposed), by reference to the table the number of years can be seen in which this percentage must continue in order to amortize the investment and pay various rates of interest on it. As said before, the ore in sight at a given rate of exhaustion can be reduced to terms of life in sight. This certain period deducted from the total term of years required gives the life which must be provided by further discovery of ore, and this can be reduced to tons or feet of extension of given ore-bodies and a tangible position arrived at. The test can be applied in this manner to the various prices which must be realized from the base metal in sight to warrant the price.

Taking the last example and assuming that the mine is equipped, and that the price is $2,000,000, the yearly return on the price is 10%. If it is desired besides amortizing or redeeming the capital to secure a return of 7% on the investment, it will be seen by reference to the table that there will be required a life of 21.6 years. As the life visible in the ore in sight is ten years, then the extensions in depth must produce ore for 11.6 years longer—1,160,000 tons. If the ore-body is 1,000 feet long and 13 feet wide, it will furnish of gold ore 1,000 tons per foot of depth; hence the ore-body must extend 1,160 feet deeper to justify the price. Mines are seldom so simple a proposition as this example. There are usually probabilities of other ore; and in the case of base metal, then variability of price and other elements must be counted. However, once the extension in depth which is necessary is determined for various assumptions of metal value, there is something tangible to consider and to weigh with the five geological weights set out in Chapter III.

The example given can be expanded to indicate not only the importance of interest and redemption in the long extension in depth required, but a matter discussed from another point of view under "Ratio of Output." If the plant on this mine were doubled and the earnings increased to 20% ($400,000 per annum) (disregarding the reduction in working expenses that must follow expansion of equipment), it will be found that the life required to repay the purchase money,—$2,000,000,—and 7% interest upon it, is about 6.8 years.

As at this increased rate of production there is in the ore in sight a life of five years, the extension in depth must be depended upon for 1.8 years, or only 360,000 tons,—that is, 360 feet of extension. Similarly, the present value of the ore in sight is $268,000 greater if the mine be given double the equipment, for thus the idle money locked in the ore is brought into the interest market at an earlier date. Against this increased profit must be weighed the increased cost of equipment. The value of low grade mines, especially, is very much a factor of the volume of output contemplated.

CHAPTER VI.

MINE VALUATION (*Concluded*).

VALUATION OF MINES WITH LITTLE OR NO ORE IN
SIGHT; VALUATIONS ON SECOND-HAND DATA;
GENERAL CONDUCT OF EXAMINATIONS; REPORTS.

A large number of examinations arise upon prospecting ventures or partially developed mines where the value is almost wholly prospective. The risks in such enterprises amount to the possible loss of the whole investment, and the possible returns must consequently be commensurate. Such business is therefore necessarily highly speculative, but not unjustifiable, as the whole history of the industry attests; but this makes the matter no easier for the mine valuer. Many devices of financial procedure assist in the limitation of the sum risked, and offer a middle course to the investor between purchase of a wholly prospective value and the loss of a possible opportunity to profit by it. The usual form is an option to buy the property after a period which permits a certain amount of development work by the purchaser before final decision as to purchase.

Aside from young mines such enterprises often arise from the possibility of lateral extension of the ore-deposit outside the boundaries of the property of original discovery (Fig. 3), in which cases there is often no visible ore within the property under consideration upon which to found opinion. In regions where vertical side lines obtain, there is always the possibility of a "deep level" in inclined deposits. Therefore the ground surrounding known deposits has a certain speculative value, upon which engineers are often called to pass judgment. Except in such unusual occurrences as South African bankets, or Lake Superior coppers, prospecting for deep level of extension is also a highly speculative phase of mining.

The whole basis of opinion in both classes of ventures must be the few geological weights,—the geology of the property and the district, the development of surrounding mines, etc. In any event, there is a very great percentage of risk, and the profit to be gained by success must be, proportionally to the expenditure involved, very large. It is no case for calculating amortization and other refinements. It is one where several hundreds or thousands of per cent hoped for on the investment is the only justification.

OPINIONS AND VALUATIONS UPON SECOND-HAND DATA.

Some one may come forward and deprecate the bare suggestion of an engineer's offering an opinion when he cannot have proper first-hand data. But in these days we have to deal with conditions as well as theories of professional ethics. The growing ownership of mines by companies, that is by corporations composed of many individuals, and with their stocks often dealt in on the public exchanges, has resulted in holders whose interest is not large enough to warrant their undertaking the cost of exhaustive examinations. The system has produced an increasing class of mining speculators and investors who are finding and supplying the enormous sums required to work our mines,—sums beyond the reach of the old-class single-handed mining men. Every year the mining investors of the new order are coming more and more to the engineer for advice, and they should be encouraged, because such counsel can be given within limits, and these limits tend to place the industry upon a sounder footing of ownership. As was said before, the lamb can be in a measure protected. The engineer's interest is to protect him, so that the industry which concerns his own life-work may be in honorable repute, and that capital may be readily forthcoming for its expansion. Moreover, by constant advice to the investor as to what constitutes a properly presented and managed project, the arrangement of such proper presentation and management will tend to become an *a priori* function of the promoter.

Sometimes the engineer can make a short visit to the mine for data purposes,—more often he cannot. In the former case, he can resolve for himself an approximation upon all the factors bearing on value, except the quality of the ore. For this, aside from inspection of the ore itself, a look at the plans is usually enlightening. A longitudinal section of the mine showing a continuous shortening of the stopes with each succeeding level carries its own interpretation. In the main, the current record of past production and estimates of the management as to ore-reserves, etc., can be accepted in ratio to the confidence that can be placed in the men who present them. It then becomes a case of judgment of men and things, and here no rule applies.

Advice must often be given upon data alone, without inspection of the mine. Most mining data present internal evidence as to credibility. The untrustworthy and inexperienced betray themselves in their every written production. Assuming the reliability of data, the methods already discussed for weighing the ultimate value of the property can be applied. It would be possible to cite hundreds of examples of valuation based upon second-hand data. Three will, however, sufficiently illustrate. First, the R mine at Johannesburg. With the regularity of this deposit, the development done, and a study of the workings on the neighboring mines and in deeper ground, it is a not unfair assumption that the reefs will maintain size and value throughout the area. The management is sound, and all the data are given in the best manner. The life of the mine is estimated at six years, with some probabilities of further ore from low-grade sections. The annual earnings available for dividends are at the rate of about £450,000 per annum. The capital is £440,000 in £1 shares. By reference to the table on page 46 it will be seen that the present value of £450,000 spread over six years to return capital at the end of that period, and give 7% dividends in the meantime, is 4.53 x £450,000 = £2,036,500 ÷ 440,000 = £4 12s. 7d. per share. So that this mine, on the assumption of continuity of values, will pay about 7% and return the price. Seven per cent is, however, not deemed an adequate return for the risks of labor troubles, faults, dykes, or poor patches. On a 9% basis, the mine is worth about £4 4s. per share.

Second, the G mine in Nevada. It has a capital of $10,000,000 in $1 shares, standing in the market at 50 cents each. The reserves are 250,000 tons, yielding a profit for yearly division of $7 per ton. It has an annual capacity of about 100,000 tons, or $700,000 net profit, equal to 14% on the market value. In order to repay the capital value of $5,000,000 and 8% per annum, it will need a life of (Table III) 13 years, of which 2-1/2 are visible. The size of the ore-bodies indicates a yield of about 1,100 tons per foot of depth. At an exhaustion rate of 100,000 tons per annum, the mine would need to extend to a depth of over a thousand feet below the present bottom. There is always a possibility of finding parallel bodies or larger volumes in depth, but it would be a sanguine engineer indeed who would recommend the stock, even though it pays an apparent 14%.

Third, the B mine, with a capital of $10,000,000 in 2,000,000 shares of $5 each. The promoters state that the mine is in the slopes of the Andes in Peru; that there are 6,000,000 tons of "ore blocked out"; that two assays by the assayers of the Bank of England average 9% copper; that the copper can be produced at five cents per pound; that there is thus a profit of $10,000,000 in sight. The evidences are wholly incompetent. It is a gamble on statements of persons who have not the remotest idea of sound mining.

GENERAL CONDUCT OF EXAMINATION.

Complete and exhaustive examination, entailing extensive sampling, assaying, and metallurgical tests, is very expensive and requires time. An unfavorable report usually means to the employer absolute loss of the engineer's fee and expenses. It becomes then the initial duty of the latter to determine at once, by the general conditions surrounding the property, how far the expenditure for exhaustive examination is warranted. There is usually named a money valuation for the property, and thus a peg is afforded upon which to hang conclusions. Very often collateral factors with a preliminary sampling, or indeed no sampling at all, will determine the whole business. In fact, it is becoming very common to send younger engineers to report as to whether exhaustive examination by more expensive men is justified.

In the course of such preliminary inspection, the ore-bodies may prove to be too small to insure adequate yield on the price, even assuming continuity in depth and represented value. They may be so difficult to mine as to make costs prohibitive, or they may show strong signs of "petering out." The ore may present visible metallurgical difficulties which make it unprofitable in any event. A gold ore may contain copper or arsenic, so as to debar cyanidation, where this process is the only hope of sufficiently moderate costs. A lead ore may be an amorphous compound with zinc, and successful concentration or smelting without great penalties may be precluded. A copper ore may carry a great excess of silica and be at the same time unconcentratable, and there may be no base mineral supply available for smelting mixture. The mine may be so small or so isolated that the cost of equipment will never be justified. Some of these conditions may be determined as unsurmountable, assuming a given value for the ore, and may warrant the rejection of the mine at the price set.

It is a disagreeable thing to have a disappointed promoter heap vituperation on an engineer's head because he did not make an exhaustive examination. Although it is generally desirable to do some sampling to give assurance to both purchaser and vendor of conscientiousness, a little courage of conviction, when this is rightly and adequately grounded, usually brings its own reward.

Supposing, however, that conditions are right and that the mine is worth the price, subject to confirmation of values, the determination of these cannot be undertaken unless time and money are available for the work. As was said, a sampling campaign is expensive, and takes time, and no engineer has the moral right to undertake an examination unless both facilities are afforded. Curtailment is unjust, both to himself and to his employer.

How much time and outlay are required to properly sample a mine is obviously a question of its size, and the character of its ore. An engineer and one principal assistant can conduct two sampling parties. In hard rock it may be impossible to take more than five samples a day for each party. But, in average ore, ten samples for each is reasonable work. As the number of samples is dependent upon the footage of openings on the deposit, a rough approximation can be made in advance, and a general idea obtained as to the time required. This period must be insisted upon.

REPORTS.

Reports are to be read by the layman, and their first qualities should be simplicity of terms and definiteness of conclusions. Reports are usually too long, rather than too short. The essential facts governing the value of a mine can be expressed on one sheet of paper. It is always desirable, however, that the ground-work data and the manner of their determination should be set out with such detail that any other engineer could come to the same conclusion if he accepted the facts as accurately deter-mined. In regard to the detailed form of reports, the writer's own preference is for a single page summarizing the main factors, and an assay plan, reduced to a longitudinal section where possi-ble. Then there should be added, for purposes of record and for submission to other engineers, a set of appendices going into some details as to the history of the mine, its geology, develop-ment, equipment, metallurgy, and management. A list of samples should be given with their location, and the tonnages and values of each separate block. A presentation should be made of the probabilities of extension in depth, together with recommendations for working the mine.

GENERAL SUMMARY.

The bed-rock value which attaches to a mine is the profit to be won from proved ore and in which the price of metal is calcu-lated at some figure between "basic" and "normal." This we may call the "A" value. Beyond this there is the speculative value of the mine. If the value of the "probable" ore be represented by X, the value of extension of the ore by Y, and a higher price for metal than the price above assumed represented by Z, then if the mine be efficiently managed the value of the mine is $A + X + Y + Z$. What actual amounts should be attached to X, Y, Z is a matter of judgment. There is no prescription for good judgment. Good judgment rests upon a proper balancing of evidence. The amount of risk in X, Y, Z is purely a question of how much these factors are required to represent in money,—in effect, how much more ore must be found, or how many feet the ore must extend in depth; or in convertible terms, what life in years the mine must have, or how high the price of metal must be. In forming an opinion whether these requirements will be realized, X, Y, Z must be balanced in a scale whose measuring standards are the five geological weights and the general industrial outlook. The wise engineer will put before his clients the scale, the weights, and the conclusion arrived at. The shrewd investor will require to know these of his adviser.

CHAPTER VII.

Development OF MINES.

ENTRY TO THE MINE; TUNNELS; VERTICAL,
INCLINED, AND COMBINED SHAFTS; LOCATION AND
NUMBER OF SHAFTS.

Development is conducted for two purposes: first, to search for ore; and second, to open avenues for its extraction. Although both objects are always more or less in view, the first predominates in the early life of mines, the prospecting stage, and the second in its later life, the producing stage. It is proposed to discuss development designed to embrace extended production purposes first, because development during the prospecting stage is governed by the same principles, but is tempered by the greater degree of uncertainty as to the future of the mine, and is, therefore, of a more temporary character.

ENTRY TO THE MINE.

There are four methods of entry: by tunnel, vertical shaft, inclined shaft, or by a combination of the last two, that is, by a shaft initially vertical then turned to an incline. Combined shafts are largely a development of the past few years to meet "deep level" conditions, and have been rendered possible only by skip-winding. The angle in such shafts (Fig. 2) is now generally made on a parabolic curve, and the speed of winding is then less diminished by the bend.

The engineering problems which present themselves under "entry" may be divided into those of:—

    1.   Method.
    2.   Location.
    3.   Shape and size.

The resolution of these questions depends upon the:—
a. Degree of dip of the deposit.
b. Output of ore to be provided for.
c. Depth at which the deposit is to be attacked.
d. Boundaries of the property.
e. Surface topography.
       f. Cost.
       g. Operating efficiency.
       h. Prospects of the mine.

FIG. 2.—Showing arrangement of the bend in combined shafts.

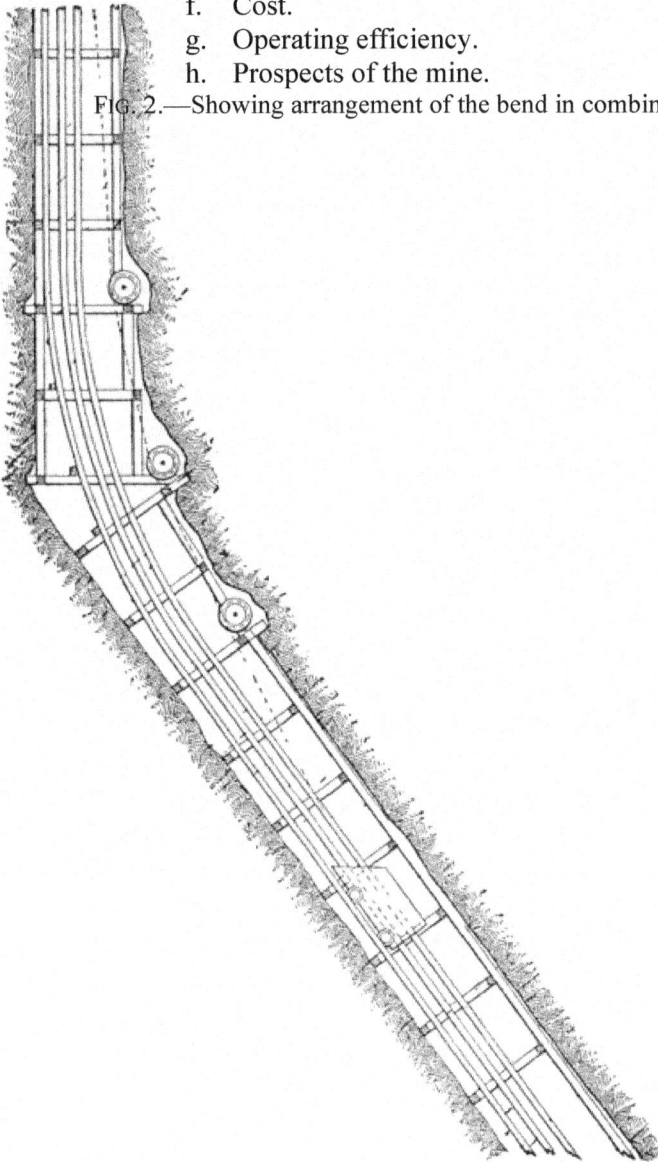

From the point of view of entrance, the co-operation of a majority of these factors permits the division of mines into certain broad classes. The type of works demanded for moderate depths (say vertically 2,500 to 3,000 feet) is very different from that required for great depths. To reach great depths, the size of shafts must greatly expand, to provide for extended ventilation, pumping, and winding necessities. Moreover inclined shafts of a degree of flatness possible for moderate depths become too long to be used economically from the surface. The vast majority of metal-mining shafts fall into the first class, those of moderate depths. Yet, as time goes on and ore-deposits are exhausted to lower planes, problems of depth will become more common. One thing, however, cannot be too much emphasized, especially on mines to be worked from the outcrop, and that is, that no engineer is warranted, owing to the speculation incidental to extension in depth, in initiating early in the mine's career shafts of such size or equipment as would be available for great depths. Moreover, the proper location of a shaft so as to work economically extension of the ore-bodies is a matter of no certainty, and therefore shafts of speculative mines are tentative in any event.

Another line of division from an engineering view is brought about by a combination of three of the factors mentioned. This is the classification into "outcrop" and "deep-level" mines. The former are those founded upon ore-deposits to be worked from or close to the surface. The latter are mines based upon the extension in depth of ore-bodies from outcrop mines. Such projects are not so common in America, where the law in most districts gives the outcrop owner the right to follow ore beyond his side-lines, as in countries where the boundaries are vertical on all sides. They do, however, arise not alone in the few American sections where the side-lines are vertical boundaries, but in other parts owing to the pitch of ore-bodies through the end lines (Fig. 3). More especially do such problems arise in America in effect, where the ingress questions have to be revised for mines worked out in the upper levels (Fig. 7). If from a standpoint of entrance questions, mines are first

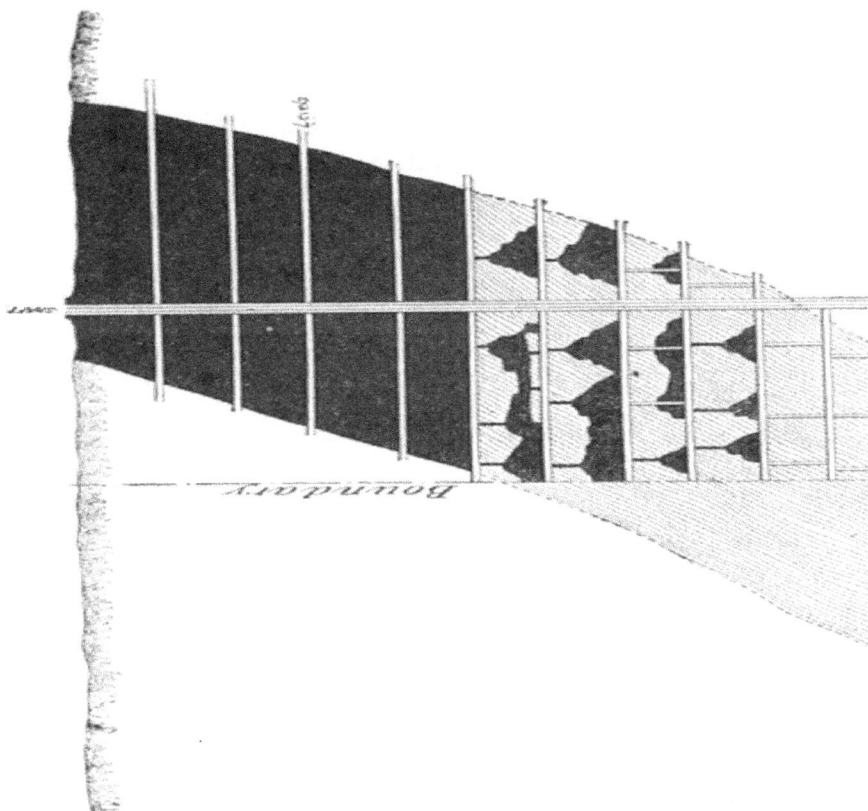

FIG.3.—Longitudinal section showing "deep level" project arising from dip of ore-body through end-line.

classified into those whose works are contemplated for moderate depths, and those in which work is contemplated for great depth, further clarity in discussion can be gained by subdivision into the possible cases arising out of the factors of location, dip, topography, and boundaries.

MINES OF MODERATE DEPTHS.

Ca se I. Deposits where topographic conditions permit the alternatives of shaft or tunnel.

Ca se II. Vertical or horizontal deposits, the only practical means of attaining which is by a vertical shaft.

Ca se III. Inclined deposits to be worked from near the surface. There are in such instances the alternatives of either a vertical or an inclined shaft.

Ca se IV. Inclined deposits which must be attacked in depth, that is, deep-level projects. There are the alternatives of a compound shaft or of a vertical shaft, and in some cases of an incline from the surface.

MINES TO GREAT DEPTHS.

Cas e V. Vertical or horizontal deposits, the only way of reaching which is by a vertical shaft.

Cas e VI. Inclined deposits. In such cases the alternatives are a vertical or a compound shaft.

Case I.—Although for logical arrangement tunnel entry has been given first place, to save repetition it is proposed to consider it later. With few exceptions, tunnels are a temporary expedient in the mine, which must sooner or later be opened by a shaft.

Case II. Vertical or Horizontal Deposits.—These require no discussion as to manner of entry. There is no justifiable alternative to a vertical shaft (Fig. 4)

Case III. Inclined Deposits which are intended to be worked from the Outcrop, or from near It (Fig. 5).—The choice of inclined or vertical shaft is dependent upon relative cost of

.

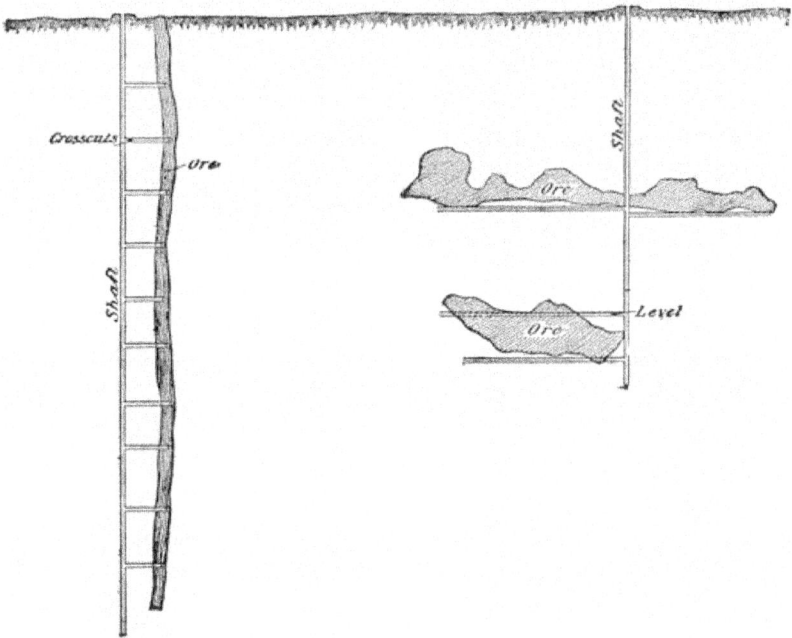

FIG.4.—Cross-sections showing entry to vertical or horizontal deposits. Case II.

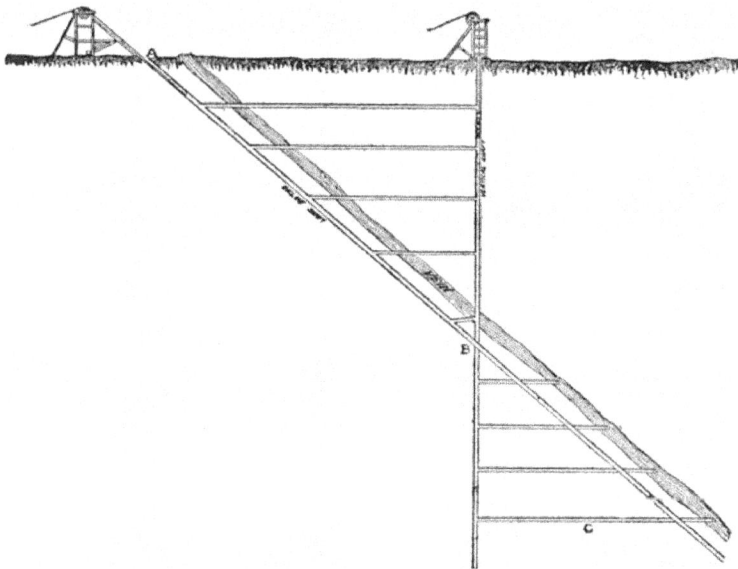

FIG.5.—Cross-section showing alternative shafts to inclined deposit to be worked from surface. Case III.

construction, subsequent operation, and the useful life of the shaft, and these matters are largely governed by the degree of dip. Assuming a shaft of the same size in either alternative, the comparative cost per foot of sinking is dependent largely on the breaking facilities of the rock under the different directions of attack. In this, the angles of the bedding or joint planes to the direction of the shaft outweigh other factors. The shaft which takes the greatest advantage of such lines of breaking weakness will be the cheapest per foot to sink. In South African experience, where inclined shafts are sunk parallel to the bedding planes of hard quartzites, the cost per foot appears to be in favor of the incline. On the other hand, sinking shafts across tight schists seems to be more advantageous than parallel to the bedding planes, and inclines following the dip cost more per foot than vertical shafts.

An inclined shaft requires more footage to reach a given point of depth, and therefore it would entail a greater total expense than a vertical shaft, assuming they cost the same per foot. The excess amount will be represented by the extra length, and this will depend upon the flatness of the dip. With vertical shafts, however, crosscuts to the deposit are necessary. In a comparative view, therefore, the cost of the crosscuts must be included with that of the vertical shaft, as they would be almost wholly saved in an incline following near the ore.

The factor of useful life for the shaft enters in deciding as to the advisability of vertical shafts on inclined deposits, from the fact that at some depth one of two alternatives has to be chosen. The vertical shaft, when it reaches a point below the deposit where the crosscuts are too long (C, Fig. 5), either becomes useless, or must be turned on an incline at the intersection with the ore (B). The first alternative means ultimately a complete loss of the shaft for working purposes. The latter has the disadvantage that the bend interferes slightly with haulage.

The following table will indicate an hypothetical extreme case,—not infrequently met. In it a vertical shaft 1,500 feet in depth is taken as cutting the deposit at the depth of 750 feet, the most favored position so far as aggregate length of crosscuts is concerned. The cost of crosscutting is taken at $20 per foot and that of sinking the vertical shaft at $75 per foot. The incline is assumed for two cases at $75 and $100 per foot respectively. The stoping height upon the ore between levels is counted at 125 feet.

DIP OF DEPOSIT FROM HORIZONTAL	DEPTH OF VERTICAL SHAFT	LENGTH OF INCLINE REQUIRED	NO. OF CROSSCUTS REQUIRED FROM V SHAFT	TOTAL LENGTH OF CROSSCUTS, FEET
80°	1,500	1,522	11	859
70°	1,500	1,595	12	1,911
60°	1,500	1,732	13	3,247
50°	1,500	1,058	15	5,389
40°	1,500	2,334	18	8,038
30°	1,500	3,000	23	16,237

COST OF CROSSCUTS $20 PER FOOT	COST VERTICAL SHAFT $75 PER FOOT	TOTAL COST OF VERTICAL AND CROSSCUTS	COST OF INCLINE $75 PER FOOT	COST OF INCLINE $100 PER FOOT
$17,180	$112,500	$129,680	$114,150	$152,200
38,220	112,500	150,720	118,625	159,500
64,940	112,500	177,440	129,900	172,230
107,780	112,500	220,280	114,850	195,800
178,760	112,500	291,260	175,050	233,400
324,740	112,500	437,240	225,000	300,000

From the above examples it will be seen that the cost of crosscuts put at ordinary level intervals rapidly outruns the extra expense of increased length of inclines. If, however, the conditions are such that crosscuts from a vertical shaft are not necessary at so frequent intervals, then in proportion to the decrease the advantages sway to the vertical shaft. Most situations wherein the crosscuts can be avoided arise in mines worked out in the upper levels and fall under Case IV, that of deep-level projects.

There can be no doubt that vertical shafts are cheaper to operate than inclines: the length of haul from a given depth is less; much higher rope speed is possible, and thus the haulage hours are less for the same output; the wear and tear on ropes, tracks, or guides is not so great, and pumping is more economical where the Cornish order of pump is used. On the other hand, with a vertical shaft must be included the cost of operating crosscuts. On mines where the volume of ore does not warrant mechanical haulage, the cost of tramming through the extra distance involved is an expense which outweighs any extra operating outlay in the inclined shaft itself. Even with mechanical haulage in crosscuts, it is doubtful if there is anything in favor of the vertical shaft on this score.

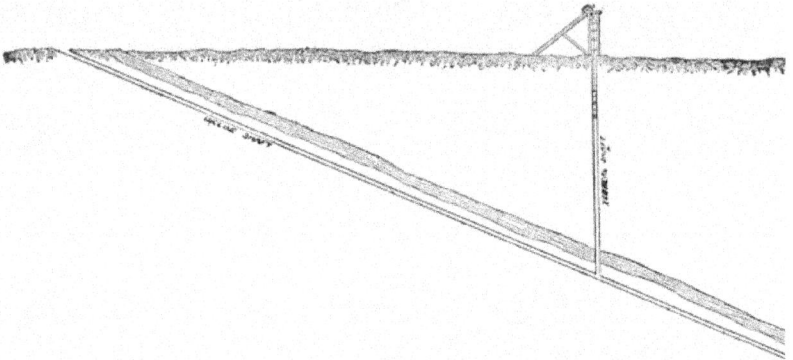

FIG. 6.—Cross-section showing auxiliary vertical outlet.

In deposits of very flat dips, under 30°, the case arises where the length of incline is so great that the saving on haulage through direct lift warrants a vertical shaft as an auxiliary outlet in addition to the incline (Fig. 6). In such a combination the crosscut question is eliminated. The mine is worked above and below the intersection by incline, and the vertical shaft becomes simply a more economical exit and an alternative to secure increased output. The North Star mine at Grass Valley is an illustration in point. Such a positive instance borders again on Case IV, deep-level projects.

In conclusion, it is the writer's belief that where mines are to be worked from near the surface, coincidentally with sinking, and where, therefore, crosscuts from a vertical shaft would need to be installed frequently, inclines are warranted in all dips under 75° and over 30°. Beyond 75° the best alternative is often undeterminable. In the range under 30° and over 15°, although inclines are primarily necessary for actual delivery of ore from levels, they can often be justifiably supplemented by a vertical shaft as a relief to a long haul. In dips of less than 15°, as in those over 75°, the advantages again trend strongly in favor of the vertical shaft. There arise, however, in mountainous countries, topographic conditions such as the dip of deposits into the mountain, which preclude any alternative on an incline at any angled dip.

Case IV. Inclined Deposits which must be attacked in Depth(Fig. 7).—There are two principal conditions in which such properties exist: first, mines being operated, or having been previously worked, whose method of entry must be revised; second, those whose ore-bodies to be attacked do not outcrop within the property.

The first situation may occur in mines of inadequate shaft capacity or wrong location; in mines abandoned and resurrected; in mines where a vertical shaft has reached its limit of useful extensions, having passed the place of economical crosscutting; or in mines in flat deposits with inclines whose haul has become too long to be economical. Three alternatives present themselves in such cases: a new incline from the surface (*A B F*, Fig. 7), or a vertical shaft combined with incline extension (*C D F*), or a simple vertical shaft (*H G*). A comparison can be first made between the simple incline and the combined shaft. The construction of an incline from the surface to the ore-body will be more costly than a combined shaft, for until the horizon of the ore is reached (at *D*) no crosscuts are required in the vertical section, while the incline must be of greater length to reach the same horizon. The case arises, however, where inclines can be sunk through old stopes, and thus more cheaply constructed than vertical shafts through solid rock; and also the case of mountainous topographic conditions mentioned above.

From an operating point of view, the bend in combined shafts (at *D*) gives rise to a good deal of wear and tear on ropes and gear. The possible speed of winding through a combined shaft is, however, greater than a simple incline, for although haulage speed through

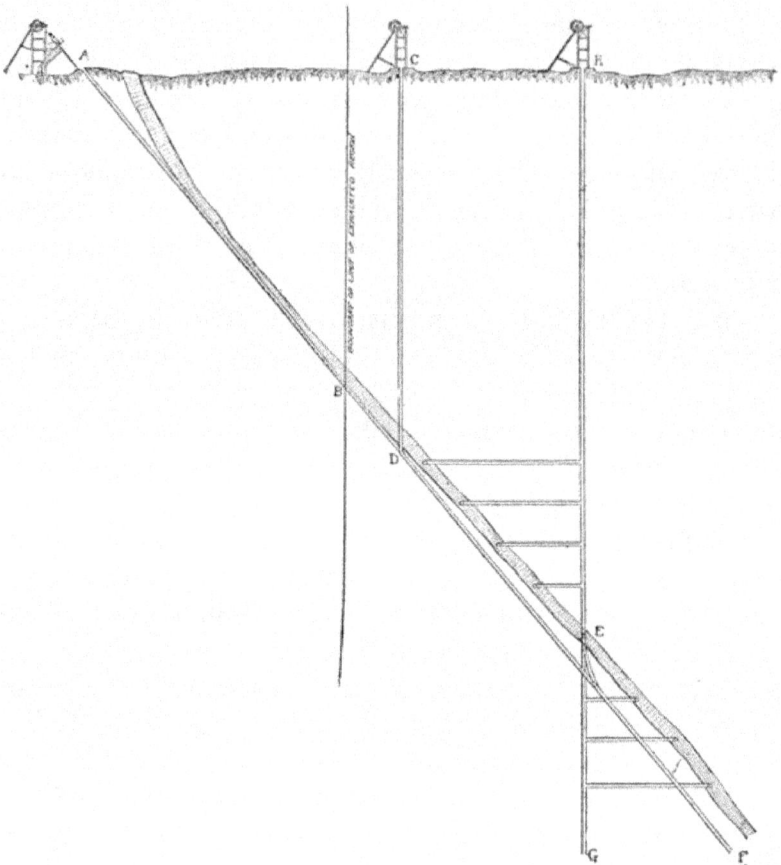

FIG.7.—Cross-section of inclined deposit which must be attacked in depth.

the incline section (*D F*) and around the bend of the combined shaft is about the same as throughout a simple incline (*A F*), the speed can be accelerated in the vertical portion (*D C*) above that feasible did the incline extend to the surface. There is therefore an advantage in this regard in the combined shaft. The net advantages of the combined over the inclined shaft depend on the comparative length of the two alternative routes from the intersection (*D*) to the surface. Certainly it is not advisable to sink a combined shaft to cut a deposit at 300 feet in depth if a simple incline can be had to the surface. On the other hand, a combined shaft cutting the deposit at 1,000 feet will be more advisable than a simple incline 2,000 feet long to reach the same point. The matter is one for direct calculation in each special case. In general, there are few instances of really deep-level projects where a complete incline from the surface is warranted.

In most situations of this sort, and in all of the second type (where the outcrop is outside the property), actual choice usually lies between combined shafts (*C D F*) and entire vertical shafts (*H G*). The difference between a combined shaft and a direct vertical shaft can be reduced to a comparison of the combined shaft below the point of intersection (*D*) with that portion of a vertical shaft which would cover the same horizon. The question then becomes identical with that of inclined *versus* verticals, as stated in Case III, with the offsetting disadvantage of the bend in the combined shaft. If it is desired to reach production at the earliest date, the lower section of a simple vertical shaft must have crosscuts to reach the ore lying above the horizon of its intersection (*E*). If production does not press, the ore above the intersection (*EB*) can be worked by rises from the horizon of intersection (*E*). In the use of rises, however, there follow the difficulties of ventilation and lowering the ore down to the shaft, which brings expenses to much the same thing as operating through crosscuts.

The advantages of combined over simple vertical shafts are earlier production, saving of either rises or crosscuts, and the ultimate utility of the shaft to any depth. The disadvantages are the cost of the extra length of the inclined section, slower winding, and greater wear and tear within the inclined section and especially around the bend. All these factors are of variable import, depending upon the dip. On very steep dips,—over 70°,—the net result is in favor of the simple vertical shaft. On other dips it is in favor of the combined shaft.

Cases V and VI. Mines to be worked to Great Depths,— over 3,000 Feet.—In Case V, with vertical or horizontal deposits, there is obviously no desirable alternative to vertical shafts.

In Case VI, with inclined deposits, there are the alternatives of a combined or of a simple vertical shaft. A vertical shaft in locations (*H*, Fig. 7) such as would not necessitate extension in depth by an incline, would, as in Case IV, compel either cross-cuts to the ore or inclines up from the horizon of intersection (*E*). Apart from delay in coming to production and the consequent loss of interest on capital, the ventilation problems with this arrangement would be appalling. Moreover, the combined shaft, entering the deposit near its shallowest point, offers the possibility of a separate haulage system on the inclined and on the vertical sections, and such separate haulage is usually advisable at great depths. In such instances, the output to be handled is large, for no mine of small output is likely to be contemplated at such depth. Several moderate-sized inclines from the horizon of intersection have been suggested (*EF*, *DG*, *CH*, Fig. 8) to feed a large primary shaft (*AB*), which thus becomes the trunk road. This program would cheapen lateral haulage underground, as mechanical traction can be used in the main level, (*EC*), and horizontal haulage costs can be reduced on the lower levels. Moreover, separate winding engines on the two sections increase the capacity, for the effect is that of two trains instead of one running on a single track.

Shaft Location.—Although the prime purpose in locating a shaft is obviously to gain access to the largest volume of ore within the shortest haulage distance, other conditions also enter, such as the character of the surface and the rock to be intersected, the time involved before reaching production, and capital cost. As shafts must bear two relations to a deposit,—one as to the dip and the other as to the strike,—they may be considered from these aspects. Vertical shafts must be on the hanging-wall side of the outcrop if the deposit dips at all. In any event, the shaft should be far enough away to be out of the reach of creeps. An inclined shaft may be sunk either on the vein, in which case a pillar of ore must be left to support the shaft; or, instead, it may be sunk a short distance in the footwall, and where necessary the excavation above can be supported by filling. Following the ore has the advantage of prospecting in sinking, and in many cases the softness of the ground in the region

of the vein warrants this procedure. It has, however, the disadvantage that a pillar of ore is locked up until the shaft is ready
for abandonment. Moreover, as veins or lodes are seldom of
even dip, an inclined shaft, to have value as a prospecting

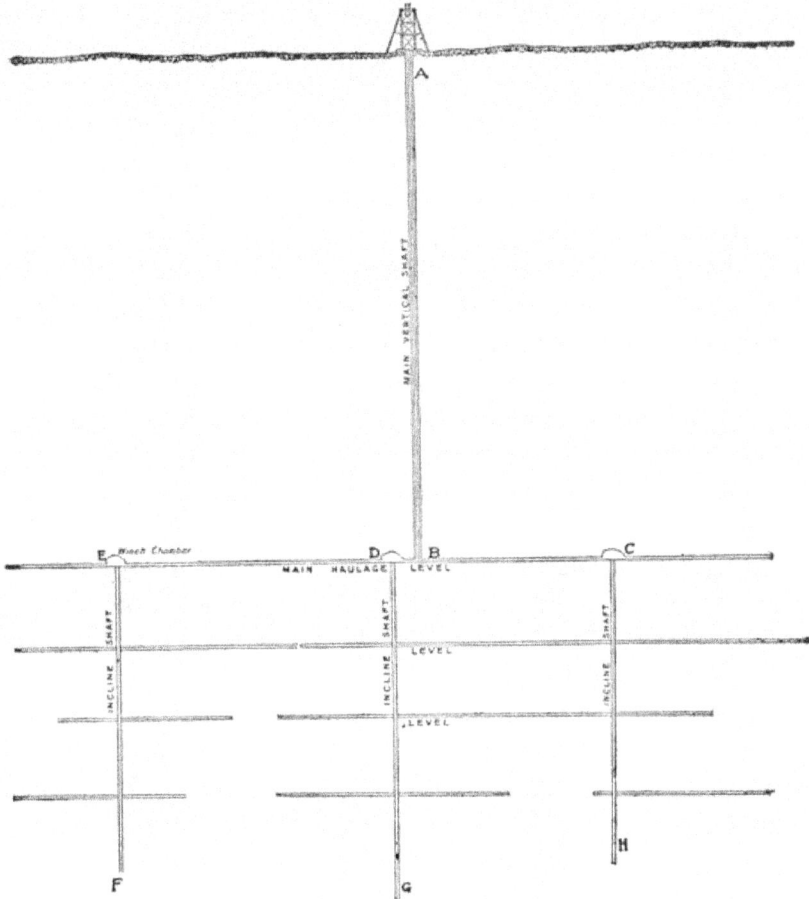

FIG.8.—Longitudinal section showing shaft arrangement proposed for
very deep inclined deposits.

opening, or to take advantage of breaking possibilities in the
lode, will usually be crooked, and an incline irregular in detail
adds greatly to the cost of winding and maintenance. These twin
disadvantages usually warrant a straight incline in the footwall.
Inclines are not necessarily of the same dip throughout, but for
reasonably economical haulage change of angle must take place
gradually.

In the case of deep-level projects on inclined deposits, demanding combined or vertical shafts, the first desideratum is to locate the vertical section as far from the outcrop as possible, and thus secure the most ore above the horizon of intersection. This, however, as stated before, would involve the cost of crosscuts or rises and would cause delay in production, together with the accumulation of capital charges. How important the increment of interest on capital may become during the period of opening the mine may be demonstrated by a concrete case. For instance, the capital of a company or the cost of the property is, say, $1,000,000, and where opening the mine for production requires four years, the aggregate sum of accumulated compound interest at 5% (and most operators want more from a mining investment) would be $216,000. Under such circumstances, if a year or two can be saved in getting to production by entering the property at a higher horizon, the difference in accumulated interest will more than repay the infinitesimal extra cost of winding through a combined shaft of somewhat increased length in the inclined section.

The unknown character of the ore in depth is always a sound reason for reaching it as quickly and as cheaply as possible. In result, such shafts are usually best located when the vertical section enters the upper portion of the deposit.

The objective in location with regard to the strike of the ore-bodies is obviously to have an equal length of lateral ore-haul in every direction from the shaft. It is easier to specify than to achieve this, for in all speculative deposits ore-shoots are found to pursue curious vagaries as they go down. Ore-bodies do not reoccur with the same locus as in the upper levels, and generally the chances to go wrong are more numerous than those to go right.

Number of Shafts.—The problem of whether the mine is to be opened by one or by two shafts of course influences location. In metal mines under Cases II and III (outcrop properties) the ore output requirements are seldom beyond the capacity of one shaft. Ventilation and escape-ways are usually easily managed through the old stopes. Under such circumstances, the conditions warranting a second shaft are the length of underground haul and isolation of ore-bodies or veins. Lateral haulage underground is necessarily disintegrated by the various levels, and usually has to be done by hand. By shortening this distance of tramming and by consolidation of the material from all levels at the surface, where mechanical haulage can be installed, a second shaft is often justified. There is therefore an economic limitation to the radius of a single shaft, regardless of the ability of the shaft to handle the total output.

Other questions also often arise which are of equal importance to haulage costs. Separate ore-shoots or ore-bodies or parallel deposits necessitate, if worked from one shaft, constant levels through unpayable ground and extra haul as well, or ore-bodies may dip away from the original shaft along the strike of the deposit and a long haulage through dead levels must follow. For instance, levels and crosscuts cost roughly one-quarter as much per foot as shafts. Therefore four levels in barren ground, to reach a parallel vein or isolated ore-body 1,000 feet away, would pay for a shaft 1,000 feet deep. At a depth of 1,000 feet, at least six levels might be necessary. The tramming of ore by hand through such a distance would cost about double the amount to hoist it through a shaft and transport it mechanically to the dressing plant at surface. The aggregate cost and operation of barren levels therefore soon pays for a second shaft. If two or more shafts are in question, they must obviously be set so as to best divide the work.

Under Cases IV, V, and VI,—that is, deep-level projects,— ventilation and escape become most important considerations. Even where the volume of ore is within the capacity of a single shaft, another usually becomes a necessity for these reasons. Their location is affected not only by the locus of the ore, but, as said, by the time required to reach it. Where two shafts are to be sunk to inclined deposits, it is usual to set one so as to intersect the deposit at a lower point than the other. Production can be started from the shallower, before the second is entirely ready. The ore above the horizon of intersection of the deeper shaft is thus accessible from the shallower shaft, and the difficulty of long rises or crosscuts from that deepest shaft does not arise.

CHAPTER VIII.

DEVELOPMENT OF MINES (*Continued*).

SHAPE AND SIZE OF SHAFTS; SPEED OF SINKING; TUNNELS.

Shape of Shafts.—Shafts may be round or rectangular.[1] Round vertical shafts are largely applied to coal-mines, and some engineers have advocated their usefulness to the mining of the metals under discussion. Their great advantages lie in their structural strength, in the large amount of free space for ventilation, and in the fact that if walled with stone, brick, concrete, or steel, they can be made water-tight so as to prevent inflow from water-bearing strata, even when under great pressure. The round walled shafts have a longer life than timbered shafts. All these advantages pertain much more to mining coal or iron than metals, for unsound, wet ground is often the accompaniment of coal-measures, and seldom troubles metal-mines. Ventilation requirements are also much greater in coal-mines. From a metal-miner's standpoint, round shafts are comparatively much more expensive than the rectangular timbered type.[2] For a larger area must be excavated for the same useful space, and if support is needed, satisfactory walling, which of necessity must be brick, stone, concrete, or steel, cannot be cheaply accomplished under the conditions prevailing in most metal regions. Although such shafts would have a longer life, the duration of timbered shafts is sufficient for most metal mines. It follows that, as timber is the cheapest and all things considered the most advantageous means of shaft support for the comparatively temporary character of metal mines, to get the strains applied to the timbers in the

[1] Octagonal shafts were sunk in Mexico in former times. At each face of the octagon was a whim run by mules, and hauling leather buckets.
[2] The economic situation is rapidly arising in a number of localities that steel beams can be usefully used instead of timber. The same arguments apply to this type of support that apply to timber.

best manner, and to use the minimum amount of it consistent with security, and to lose the least working space, the shaft must be constructed on rectangular lines.

The variations in timbered shaft design arise from the possible arrangement of compartments. Many combinations can be imagined, of which Figures 9, 10, 11, 12, 13, and 14 are examples.

FIG. 9.

FIG. 10.

FIG. 11.

FIG. 12.

FIG. 13.

FIG. 14.

The arrangement of compartments shown in Figures 9, 10, 11, and 13 gives the greatest strength. It permits timbering to the best advantage, and avoids the danger underground involved in crossing one compartment to reach another. It is therefore generally adopted. Any other arrangement would obviously be impossible in inclined or combined shafts.

Size of Shafts.—In considering the size of shafts to be installed, many factors are involved. They are in the main:—

a. Amount of ore to be handled.
b. Winding plant.
c. Vehicle of transport.
d. Depth.
e. Number of men to be worked underground.
f. Amount of water.
g. Ventilation.
h. Character of the ground.
i. Capital outlay.
j. Operating expense.

It is not to be assumed that these factors have been stated in the order of relative importance. More or less emphasis will be attached to particular factors by different engineers, and under different circumstances. It is not possible to suggest any arbitrary standard for calculating their relative weight, and they are so interdependent as to preclude separate discussion. The usual result is a compromise between the demands of all.

Certain factors, however, dictate a minimum position, which may be considered as a datum from which to start consideration.

First, a winding engine, in order to work with any economy, must be balanced, that is, a descending empty skip or cage must assist in pulling up a loaded one. Therefore, except in mines of very small output, at least two compartments must be made for hoisting purposes. Water has to be pumped from most mines, escape-ways are necessary, together with room for wires and air-pipes, so that at least one more compartment must be provided for these objects. We have thus three compartments as a sound minimum for any shaft where more than trivial output is required.

Second, there is a certain minimum size of shaft excavation below which there is very little economy in actual rock-breaking.[1]

---

[1] Notes on the cost of shafts in various regions which have been personally collected show a remarkable decrease in the cost per cubic foot of material excavated with

In too confined a space, holes cannot be placed to advantage for the blast, men cannot get round expeditiously, and spoil cannot be handled readily. The writer's own experience leads him to believe that, in so far as rock-breaking is concerned, to sink a shaft fourteen to sixteen feet long by six to seven feet wide outside the timbers, is as cheap as to drive any smaller size within the realm of consideration, and is more rapid. This size of excavation permits of three compartments, each about four to five feet inside the timbers.

The cost of timber, it is true, is a factor of the size of shaft, but the labor of timbering does not increase in the same ratio. In any event, the cost of timber is only about 15% of the actual shaft cost, even in localities of extremely high prices.

Third, three reasons are rapidly making the self-dumping skip the almost universal shaft-vehicle, instead of the old cage for cars. First, there is a great economy in labor for loading into and discharging from a shaft; second, there is more rapid despatch and discharge and therefore a larger number of possible trips; third, shaft-haulage is then independent of delays in arrival of cars at stations, while tramming can be done at any time and shaft-haulage can be concentrated into certain hours. Cages to carry mine cars and handle the same load as a skip must either be big enough to take two cars, which compels a much larger shaft than is necessary with skips, or they must be double-decked, which renders loading arrangements underground costly to install and expensive to work. For all these reasons, cages can be justified only on metal mines of such small tonnage that time is no consideration and where the saving of men is not to be effected. In compartments of the minimum size mentioned above

increased size of shaft. Variations in skill, in economic conditions, and in method of accounting make data regarding different shafts of doubtful value, but the following are   of   interest:— In Australia, eight shafts between 10 and 11 feet long by 4 to 5 feet wide cost an average of $1.20 per cubic foot of material excavated. Six shafts 13 to 14 feet long by 4 to 5 feet wide cost an average of $0.95 per cubic foot; seven shafts 14 to 16 feet long and 5 to 7 feet wide cost an average of $0.82 per cubic foot. In South Africa, eleven shafts 18 to 19 feet long by 7 to 8 feet wide cost an average of $0.82 per cubic foot; five shafts 21 to 25 feet long by 8 feet wide, cost $0.74; and seven shafts 28 feet by 8 feet cost $0.60 per cubic foot.

(four to five feet either way) a skip with a capacity of from two to five tons can be installed, although from two to three tons is the present rule. Lighter loads than this involve more trips, and thus less hourly capacity, and, on the other hand, heavier loads require more costly engines. This matter is further discussed under "Haulage Appliances."

We have therefore as the economic minimum a shaft of three compartments (Fig. 9), each four to five feet square. When the maximum tonnage is wanted from such a shaft at the least operating cost, it should be equipped with loading bins and skips.

The output capacity of shafts of this size and equipment will depend in a major degree upon the engine employed, and in a less degree upon the hauling depth. The reason why depth is a subsidiary factor is that the rapidity with which a load can be drawn is not wholly a factor of depth. The time consumed in hoisting is partially expended in loading, in acceleration and retardation of the engine, and in discharge of the load. These factors are constant for any depth, and extra distance is therefore accomplished at full speed of the engine.

Vertical shafts will, other things being equal, have greater capacity than inclines, as winding will be much faster and length of haul less for same depth. Since engines have, however, a great tractive ability on inclines, by an increase in the size of skip it is usually possible partially to equalize matters. Therefore the size of inclines for the same output need not differ materially from vertical shafts.

The maximum capacity of a shaft whose equipment is of the character and size given above, will, as stated, decrease somewhat with extension in depth of the haulage horizon. At 500 feet, such a shaft if vertical could produce 70 to 80 tons per hour comfortably with an engine whose winding speed was 700 feet per minute. As men and material other than ore have to be handled in and out of the mine, and shaft-sinking has to be attended to, the winding engine cannot be employed all the time on ore. Twelve hours of actual daily ore-winding are all that can be expected without auxiliary help. This represents a capacity from such a depth of 800 to 1,000 tons per day. A similar shaft, under ordinary working conditions, with an engine speed of 2,000 feet per minute, should from, say, 3,000 feet have a capacity of about 400 to 600 tons daily.

It is desirable to inquire at what stages the size of shaft should logically be enlarged in order to attain greater capacity. A considerable measure of increase can be obtained by relieving the main hoisting engine of all or part of its collateral duties. Where the pumping machinery is not elaborate, it is often possible to get a small single winding compartment into the gangway without materially increasing the size of the shaft if the haulage compartments be made somewhat narrower (Fig. 10). Such a compartment would be operated by an auxiliary engine for sinking, handling tools and material, and assisting in handling men. If this arrangement can be effected, the productive time of the main engine can be expanded to about twenty hours with an addition of about two-thirds to the output.

Where the exigencies of pump and gangway require more than two and one-half feet of shaft length, the next stage of expansion becomes four full-sized compartments (Fig. 11). By thus enlarging the auxiliary winding space, some assistance may be given to ore-haulage in case of necessity. The mine whose output demands such haulage provisions can usually stand another foot of width to the shaft, so that the dimensions come to about 21 feet to 22 feet by 7 feet to 8 feet outside the timbers. Such a shaft, with three- to four-ton skips and an appropriate engine, will handle up to 250 tons per hour from a depth of 1,000 feet.

The next logical step in advance is the shaft of five compartments with four full-sized haulage ways (Fig. 13), each of greater size than in the above instance. In this case, the auxiliary engine becomes a balanced one, and can be employed part of the time upon ore-haulage. Such a shaft will be about 26 feet to 28 feet long by 8 feet wide outside the timbers, when provision is made for one gangway. The capacity of such shafts can be up to 4,000 tons a day, depending on the depth and engine. When very large quantities of water are to be dealt with and rod-driven pumps to be used, two pumping compartments are sometimes necessary, but other forms of pumps do not require more than one compartment,—an additional reason for their use.

For depths greater than 3,000 feet, other factors come into play. Ventilation questions become of more import. The mechanical problems on engines and ropes become involved, and their sum-effect is to demand much increased size and a greater number of compartments. The shafts at Johannesburg intended as outlets for workings 5,000 feet deep are as much as 46 feet by 9 feet outside timbers.

It is not purposed to go into details as to sinking methods or timbering. While important matters, they would unduly prolong this discussion. Besides, a multitude of treatises exist on these subjects and cover all the minutiæ of such work.

Speed of Sinking.—Mines may be divided into two cases,— those being developed only, and those being operated as well as developed. In the former, the entrance into production is usually dependent upon the speed at which the shaft is sunk. Until the mine is earning profits, there is a loss of interest on the capital involved, which, in ninety-nine instances out of a hundred, warrants any reasonable extra expenditure to induce more rapid progress. In the case of mines in operation, the volume of ore available to treatment or valuation is generally dependent to a great degree upon the rapidity of the extension of workings in depth. It will be demonstrated later that, both from a financial and a technical standpoint, the maximum development is the right one and that unremitting extension in depth is not only justifiable but necessary.

Speed under special conditions or over short periods has a more romantic than practical interest, outside of its value as a stimulant to emulation. The thing that counts is the speed which can be maintained over the year. Rapidity of sinking depends mainly on:—

  a. Whether the shaft is or is not in use for operating the mine.
  b. The breaking character of the rock.
  c. The amount of water.

The delays incident to general carrying of ore and men are such that the use of the main haulage engine for shaft-sinking is practically impossible, except on mines with small tonnage output. Even with a separate winch or auxiliary winding-engine, delays are unavoidable in a working shaft, especially as it usually has more water to contend with than one not in use for operating the mine. The writer's own impression is that an average of 40 feet per month is the maximum possibility for year in and out sinking under such conditions. In fact, few going mines manage more than 400 feet a year. In cases of clean shaft-sinking, where every energy is bent to speed, 150 feet per month have been averaged for many months. Special cases have occurred where as much as 213 feet have been achieved in a single month. With ordinary conditions, 1,200 feet in a year is very good work. Rock awkward to break, and water especially, lowers the rate of progress very materially. Further reference to speed will be found in the chapter on "Drilling Methods."

Tunnel Entry.—The alternative of entry to a mine by tunnel is usually not a question of topography altogether, but, like everything else in mining science, has to be tempered to meet the capital available and the expenditure warranted by the value showing.

In the initial prospecting of a mine, tunnels are occasionally overdone by prospectors. Often more would be proved by a few inclines. As the pioneer has to rely upon his right arm for hoisting and drainage, the tunnel offers great temptations, even when it is long and gains but little depth. At a more advanced stage of development, the saving of capital outlay on hoisting and pumping equipment, at a time when capital is costly to secure, is often sufficient justification for a tunnel entry. But at the stage where the future working of ore below a tunnel-level must be contemplated, other factors enter. For ore below tunnel-level a shaft becomes necessary, and in cases where a tunnel enters a few hundred feet below the outcrop the shaft should very often extend to the surface, because internal shafts, winding from tunnel-level, require large excavations to make room for the transfer of ore and for winding gear. The latter must be operated by transmitted power, either that of steam, water, electricity, or air. Where power has to be generated on the mine, the saving by the use of direct steam, generated at the winding gear, is very considerable. Moreover, the cost of haulage through a shaft for the extra distance from tunnel-level to the surface is often less than the cost of transferring the ore and removing it through the tunnel. The load once on the winding-engine, the consumption of power is small for the extra distance, and the saving of labor is of consequence. On the other hand, where drainage problems arise, they usually outweigh all other considerations, for whatever the horizon entered by tunnel, the distance from that level to the surface means a saving of water-pumpage against so much head. The accumulation of such constant expense justifies a proportioned capital outlay. In other words, the saving of this extra pumping will annually redeem the cost of a certain amount of tunnel, even though it be used for drainage only.

In order to emphasize the rapidity with which such a saving of constant expense will justify capital outlay, one may tabulate the result of calculations showing the length of tunnel warranted with various hypothetical factors of quantity of water and height of lift eliminated from pumping. In these computations, power is taken at the low rate of $60 per horsepower-year, the cost of tunneling at an average figure of $20 per foot, and the time on the basis of a ten-year life for the mine.

FEET OF TUNNEL PAID FOR IN 10 YEARS WITH UNDER-MENTIONED CONDITIONS.

FEET OF WATER LIFT AVOIDED	100,000 GALLONS PER DIEM	200,000 GALLONS PER DIEM	300,000 GALLONS PER DIEM	500,000 GALLONS PER DIEM	1,000,000 GALLONS PER DIEM
100	600	1,200	1,800	3,000	6,000
200	1,200	2,400	3,600	6,000	12,000
300	1,800	3,600	5,400	9,000	18,000
500	3,000	6,000	9,000	15,000	30,000
1,000	6,000	12,000	18,000	30,000	60,000

The size of tunnels where ore-extraction is involved depends upon the daily tonnage output required, and the length of haul. The smallest size that can be economically driven and managed is about 6-1/2 feet by 6 feet inside the timbers. Such a tunnel, with single track for a length of 1,000 feet, with one turnout, permits handling up to 500 tons a day with men and animals. If the distance be longer or the tonnage greater, a double track is required, which necessitates a tunnel at least 8 feet wide by 6-1/2 feet to 7 feet high, inside the timbers.

There are tunnel projects of a much more impressive order than those designed to operate upper levels of mines; that is, long crosscut tunnels designed to drain and operate mines at very considerable depths, such as the Sutro tunnel at Virginia City. The advantage of these tunnels is very great, especially for drainage, and they must be constructed of large size and equipped with appliances for mechanical haulage.

CHAPTER IX.

DEVELOPMENT OF MINES (*Concluded*).

SUBSIDIARY DEVELOPMENT;—STATIONS;
CROSSCUTS; LEVELS; INTERVAL BETWEEN LEVELS;
PROTECTION OF LEVELS; WINZES AND RISES.
DEVELOPMENT IN THE PROSPECTING STAGE;
DRILLING.

SUBSIDIARY DEVELOPMENT.

Stations, crosscuts, levels, winzes, and rises follow after the initial entry. They are all expensive, and the least number that will answer is the main desideratum.

Stations.—As stations are the outlets of the levels to the shaft, their size and construction is a factor of the volume and character of the work at the levels which they are to serve. If no timber is to be handled, and little ore, and this on cages, the stations need be no larger than a good sized crosscut. Where timber is to be let down, they must be ten to fifteen feet higher than the floor of the crosscut. Where loading into skips is to be provided for, bins must be cut underneath and sufficient room be provided to shift the mine cars comfortably. Such bins are built of from 50 to 500 tons' capacity in order to contain some reserve for hoisting purposes, and in many cases separate bins must be provided on opposite sides of the shaft for ore and waste. It is a strong argument in favor of skips, that with this means of haulage storage capacity at the stations is possible, and the hoisting may then go on independently of trucking and, as said before, there are no idle men at the stations.

It is always desirable to concentrate the haulage to the least number of levels, for many reasons. Among them is that, where haulage is confined to few levels, storage-bins are

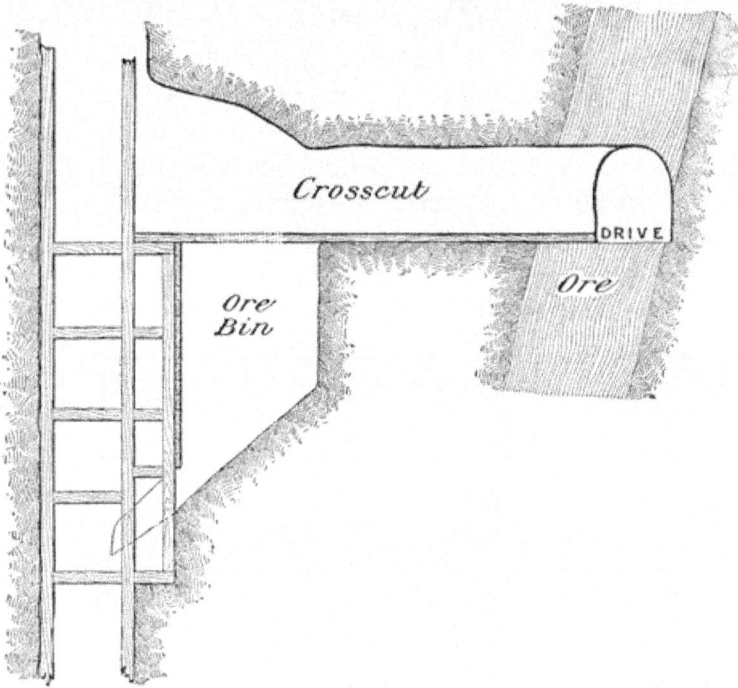

FIG.15.—Cross-section of station arrangement for skip-haulage in vertical shaft.

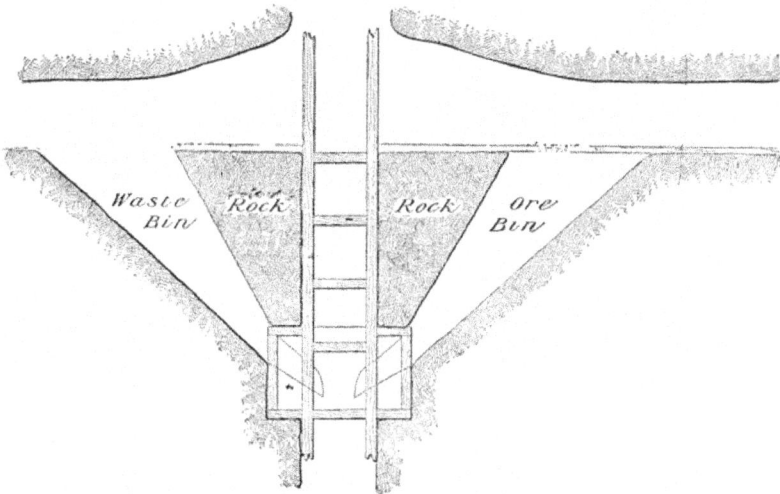

FIG.16.—Cross-section of station arrangement for skip-haulage in vertical shaft.

not required at every station. Figures 15, 16, 17, and 18 illustrate various arrangements of loading bins.

Crosscuts.—Crosscuts are for two purposes, for roadway connection of levels to the shaft or to other levels, and for prospecting purposes. The number of crosscuts for roadways can sometimes be decreased by making the connections with the shaft at every second or even every third level, thus not only saving in the construction cost of crosscuts and stations, but also in the expenses of scattered tramming. The matter becomes especially worth considering where the quantity of ore that can thus be accumulated warrants mule or mechanical haulage. This subject will be referred to later on.

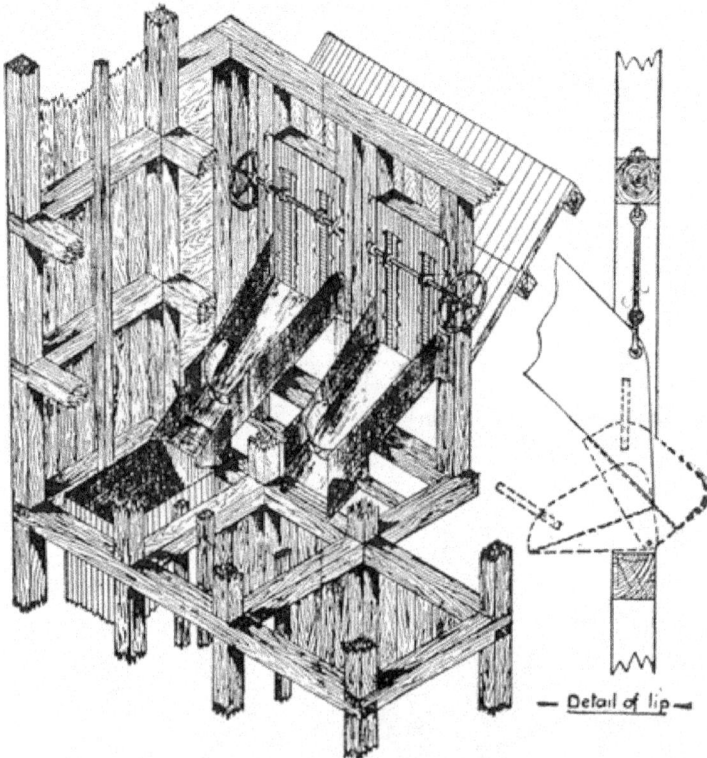

FIG.17.—Arrangement of loading chutes in vertical shaft.

On the second purpose of crosscuts,—that of prospecting,—one observation merits emphasis. This is, that the tendency of ore-fissures to be formed in parallels warrants more systematic crosscutting into the country rock than is done in many mines.

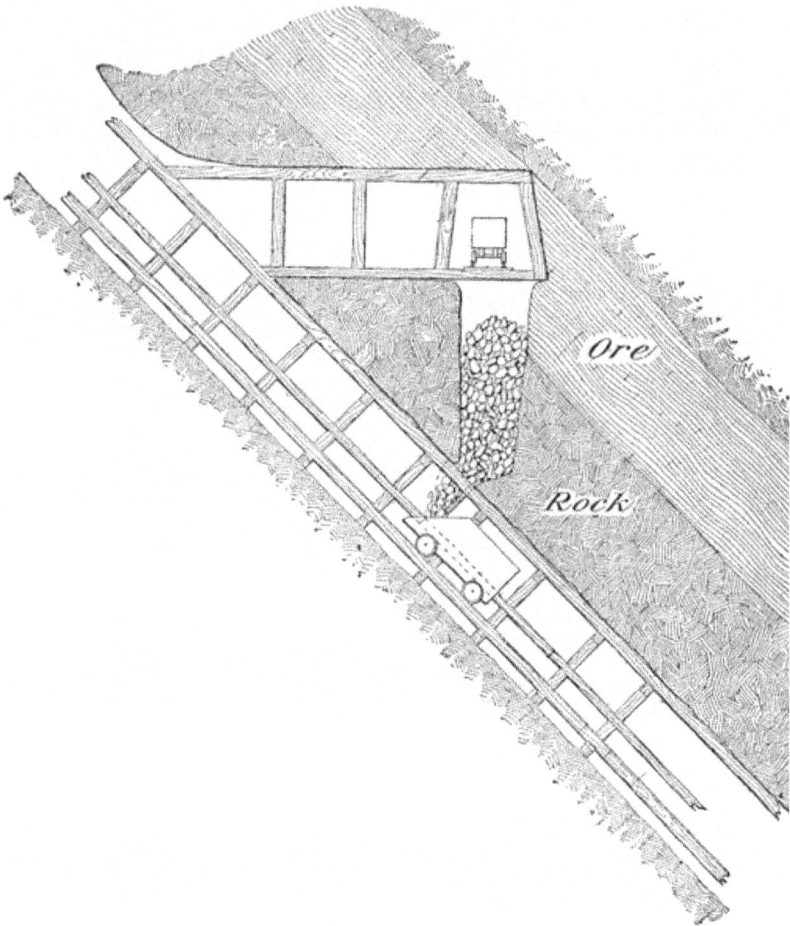

FIG.18.—Cross-section of station arrangement for skip-haulage in inclined shaft.

LEVELS.

The word "level" is another example of miners' adaptations in nomenclature. Its use in the sense of tunnels driven in the direction of the strike of the deposit has better, but less used, synonyms in the words "drifts" or "drives." The term "level" is used by miners in two senses, in that it is sometimes applied to all openings on one horizon, crosscuts included. Levels are for three purposes,—for a stoping base; for prospecting the deposit; and for roadways. As a prospecting and a stoping base it is desirable that the level should be driven on the deposit; as a roadway, that it should constitute the shortest distance between two points and be in the soundest ground. On narrow, erratic deposits the levels usually must serve all three purposes at once; but in wider and more regular deposits levels are often driven separately for roadways from the level which forms the stoping base and prospecting datum.

There was a time when mines were worked by driving the level on ore and enlarging it top and bottom as far as the ground would stand, then driving the next level 15 to 20 feet below, and repeating the operation. This interval gradually expanded, but for some reason 100 feet was for years assumed to be the proper distance between levels. Scattered over every mining camp on earth are thousands of mines opened on this empirical figure, without consideration of the reasons for it or for any other distance.

The governing factors in determining the vertical interval between levels are the following:—

a. The regularity of the deposit.

b. The effect of the method of excavation of winzes and rises.

c. The dip and the method of stoping.

Regularity of the Deposit.—From a prospecting point of view the more levels the better, and the interval therefore must be determined somewhat by the character of the deposit. In erratic deposits there is less risk of missing ore with frequent levels, but it does not follow that every level need be a through roadway to the shaft or even a stoping base. In such deposits, intermediate levels for prospecting alone are better than complete levels, each a roadway. Nor is it essential, even where frequent levels are required for a stoping base, that each should be a main haulage outlet to the shaft. In some mines every third level is used as a main roadway, the ore being poured from the intermediate ones down to the haulage line. Thus tramming and shaft work, as stated before, can be concentrated.

Effect of Method of Excavating Winzes and Rises.—With hand drilling and hoisting, winzes beyond a limited depth become very costly to pull spoil out of, and rises too high become difficult to ventilate, so that there is in such cases a limit to the interval desirable between levels, but these difficulties largely disappear where air-winches and air-drills are used.

The Dip and Method of Stoping.—The method of stoping is largely dependent upon the dip, and indirectly thus affects level intervals. In dips under that at which material will "flow" in the stopes—about 45° to 50°—the interval is greatly dependent on the method of stope-transport. Where ore is to be shoveled from stopes to the roadway, the levels must be comparatively close together. Where deposits are very flat, under 20°, and walls fairly sound, it is often possible to use a sort of long wall system of stoping and to lay tracks in the stopes with self-acting inclines to the levels. In such instances, the interval can be expanded to 250 or even 400 feet. In dips between 20° and 45°, tracks are not often possible, and either shoveling or "bumping troughs"[1] are the only help to transport. With shoveling, intervals of 100 feet[2] are most common, and with troughs the distance can be expanded up to 150 or 175 feet.

In dips of over 40° to 50°, depending on the smoothness of the foot wall, the distance can again be increased, as stope-transport is greatly simplified, since the stope materials fall out by gravity. In timbered stopes, in dips over about 45°, intervals of 150 to 200 feet are possible. In filled stopes intervals of over 150 feet present difficulties in the maintenance of ore-passes, for the wear and tear of longer use often breaks the timbers. In shrinkage-stopes, where no passes are to be maintained and few winzes put through, the interval is sometimes raised to 250 feet. The subject is further discussed under "Stoping."

Another factor bearing on level intervals is the needed

---

[1] Page 136.
[2] Intervals given are measured on the dip.

insurance of sufficient points of stoping attack to keep up a certain output. This must particularly influence the manager whose mine has but little ore in reserve.

FIG. 19.

Protection of Levels.—Until recent years, timbering and occasional walling was the only method for the support of the roof, and for forming a platform for a stoping base. Where the rock requires no support sublevels can be used as a stoping base, and timbering for such purpose avoided altogether (Figs. 38, 39, 42). In such cases the main roadway can then be driven on straight lines, either in the walls or in the ore, and used entirely for haulage. The subheading for a stoping base is driven far enough above or below the roadway (depending on whether overhand or underhand stoping is to be used) to leave a supporting pillar which is penetrated by short passes for ore. In overhand stopes, the ore is broken directly on the floor of an upper sublevel; and in underhand stopes, broken directly from the bottom of the sublevel. The method entails leaving a pillar of ore which can be recovered only with difficulty in mines where stope-support is necessary. The question of its adoption is then largely one of the comparative cost of timbering, the extra cost of the sublevel, and the net value of the ore left. In bad swelling veins, or badly crushing walls, where constant repair to timbers would be necessary, the use of a sublevel is a most useful alternative. It is especially useful with stopes to be left open or worked by shrinkage-stoping methods.

If the haulage level, however, is to be the stoping base, some protection to the roadway must be provided. There are three systems in use,—by wood stulls or sets (Figs. 19, 30, 43), by drywalling with timber caps (Fig. 35), and in some localities by steel sets. Stulls are put up in various ways, and, as their use entails the least difficulty in taking the ore out from beneath the level, they are much favored, but are applicable only in comparatively narrow deposits.

WINZES AND RISES.

These two kinds of openings for connecting two horizons in a mine differ only in their manner of construction. A winze is sunk underhand, while a rise is put up overhand. When the connection between levels is completed, a miner standing at the bottom usually refers to the opening as a rise, and when he goes to the top he calls it a winze. This confusion in terms makes it advisable to refer to all such completed openings as winzes, regardless of how they are constructed.

In actual work, even disregarding water, it costs on the average about 30% less to raise than to sink such openings, for obviously the spoil runs out or is assisted by gravity in one case, and in the other has to be shoveled and hauled up. Moreover, it is easier to follow the ore in a rise than in a winze. It usually happens, however, that in order to gain time both things are done, and for prospecting purposes sinking is necessary.

The number of winzes required depends upon the method of stoping adopted, and is mentioned under "Stoping." After stoping, the number necessary to be maintained open depends upon the necessities of ventilation, of escape, and of passageways for material to be used below. Where stopes are to be filled with waste, more winzes must be kept open than when other methods are used, and these winzes must be in sufficient alignment to permit the continuous flow of material down past the various levels. In order that the winzes should deliver timber and filling to the most advantageous points, they should, in dipping ore-bodies, be as far as possible on the hanging wall side.

DEVELOPMENT IN THE EARLY PROSPECTING STAGE.

The prime objects in the prospecting stage are to expose the ore and to learn regarding the ore-bodies something of their size, their value, metallurgical character, location, dip, strike, etc.,— so much at least as may be necessary to determine the works most suitable for their extraction or values warranting purchase. In outcrop mines there is one rule, and that is "follow the ore." Small temporary inclines following the deposit, even though they are eventually useless; are nine times out of ten justified.

In prospecting deep-level projects, it is usually necessary to layout work which can be subsequently used in operating the mine, because the depth involves works of such considerable scale, even for prospecting, that the initial outlay does not warrant any anticipation of revision. Such works have to be located and designed after a study of the general geology as disclosed in adjoining mines. Practically the only method of supplementing such information is by the use of churn- and diamond-drills.

Drilling.—Churn-drills are applicable only to comparatively shallow deposits of large volume. They have an advantage over the diamond drill in exposing a larger section and in their application to loose material; but inability to determine the exact horizon of the spoil does not lend them to narrow deposits, and in any event results are likely to be misleading from the finely ground state of the spoil. They are, however, of very great value for preliminary prospecting to shallow horizons.

Two facts in diamond-drilling have to be borne in mind: the indication of values is liable to be misleading, and the deflection of the drill is likely to carry it far away from its anticipated destination. A diamond-drill secures a small section which is sufficiently large to reveal the geology, but the values disclosed in metal mines must be accepted with reservations. The core amounts to but a little sample out of possibly large amounts of ore, which is always of variable character, and the core is most unlikely to represent the average of the deposit. Two diamond-drill holes on the Oroya Brownhill mine both passed through the ore-body. One apparently disclosed unpayable values, the other seemingly showed ore forty feet in width assaying $80 per ton. Neither was right. On the other hand, the predetermination of the location of the ore-body justified expenditure. A recent experiment at Johannesburg of placing a copper wedge in the hole at a point above the ore-body and deflecting the drill on reintroducing it, was successful in giving a second section of the ore at small expense.

The deflection of diamond-drill holes from the starting angle is almost universal. It often amounts to a considerable wandering from the intended course. The amount of such deflection varies with no seeming rule, but it is probable that it is especially affected by the angle at which stratification or lamination planes are inclined to the direction of the hole. A hole has been known to wander in a depth of 1,500 feet more than 500 feet from the point intended. Various instruments have been devised for surveying deep holes, and they should be brought into use before works are laid out on the basis of diamond-drill results, although none of the inventions are entirely satisfactory.

CHAPTER X.

STOPING.

METHODS OF ORE-BREAKING; UNDERHAND STOPES;
OVERHAND STOPES; COMBINED STOPE. VALUING
ORE IN COURSE OF BREAKING.

There is a great deal of confusion in the application of the word "stoping." It is used not only specifically to mean the actual ore-breaking, but also in a general sense to indicate all the operations of ore-breaking, support of excavations, and transportation between levels. It is used further as a noun to designate the hole left when the ore is taken out. Worse still, it is impossible to adhere to miners' terms without employing it in every sense, trusting to luck and the context to make the meaning clear.

The conditions which govern the method of stoping are in the main:—

    a.   The dip.
    b.   The width of the deposit.
    c.   The character of the walls.
    d.   The cost of materials.
    e.   The character of the ore.

Every mine, and sometimes every stope in a mine, is a problem special to itself. Any general consideration must therefore be simply an inquiry into the broad principles which govern the adaptability of special methods. A logical arrangement of discussion is difficult, if not wholly impossible, because the factors are partially interdependent and of varying importance.

For discussion the subject may be divided into:

    1. Methods of ore-breaking.
    2. Methods of supporting excavation.
    3. Methods of transport in stopes.

METHODS OF ORE-BREAKING.

The manner of actual ore-breaking is to drill and blast off slices from the block of ground under attack. As rock obviously breaks easiest when two sides are free, that is, when corners can be broken off, the detail of management for blasts is therefore to set the holes so as to preserve a corner for the next cut; and as a consequence the face of the stope shapes into a series of benches (Fig. 22),—inverted benches in the case of overhand stopes (Figs. 20, 21). The size of these benches will in a large measure depend on the depth of the holes. In wide stopes with machine-drills they vary from 7 to 10 feet; in narrow stopes with hand-holes, from two to three feet.

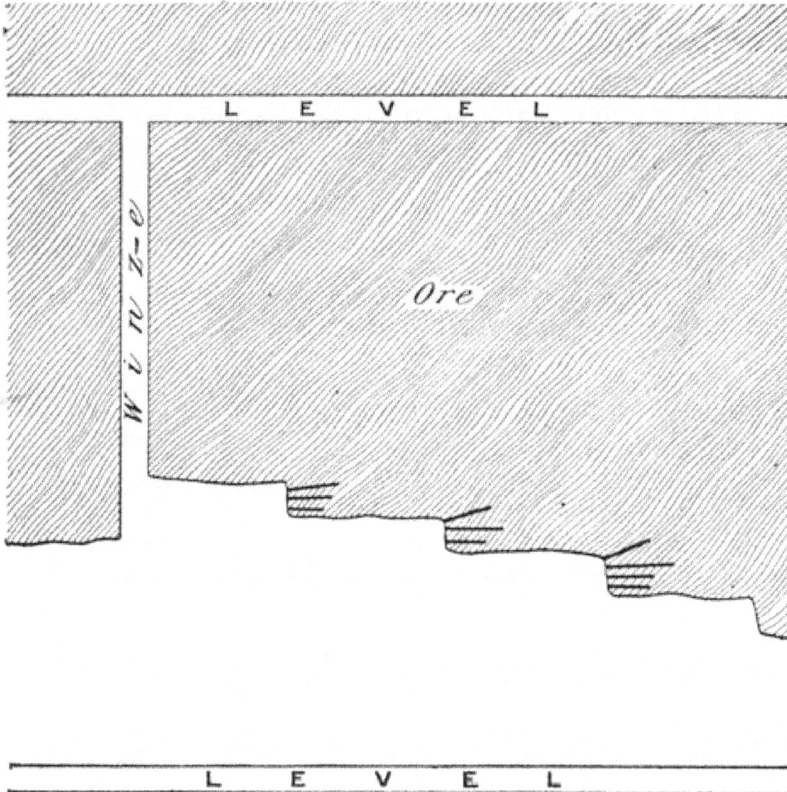

FIG. 20.

The position of the men in relation to the working face

gives rise to the usual primary classification of the methods of stoping. They are:—

1.　Underhand stopes,
2.　Overhand stopes,
3.　Combined stopes.

These terms originated from the direction of the drill-holes, but this is no longer a logical basis of distinction, for underhand holes in overhand stopes,—as in rill-stoping,—are used entirely in some mines (Fig. 21).

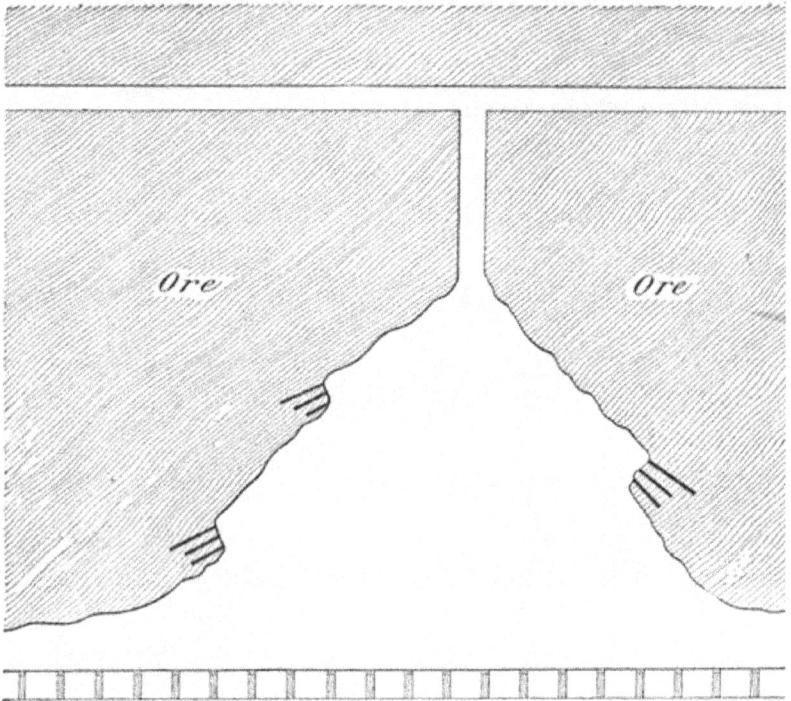

FIG. 21.

Underhand Stopes.—Underhand stopes are those in which the ore is broken downward from the levels. Inasmuch as this method has the advantage of allowing the miner to strike his blows downward and to stand upon the ore when at work, it was almost universal before the invention of powder; and was applied more generally before the invention of machine-drills than since. It is never rightly introduced unless the stope is worked back from winzes through which the ore broken can be let down to the level below, as shown in Figures 22 and 23.

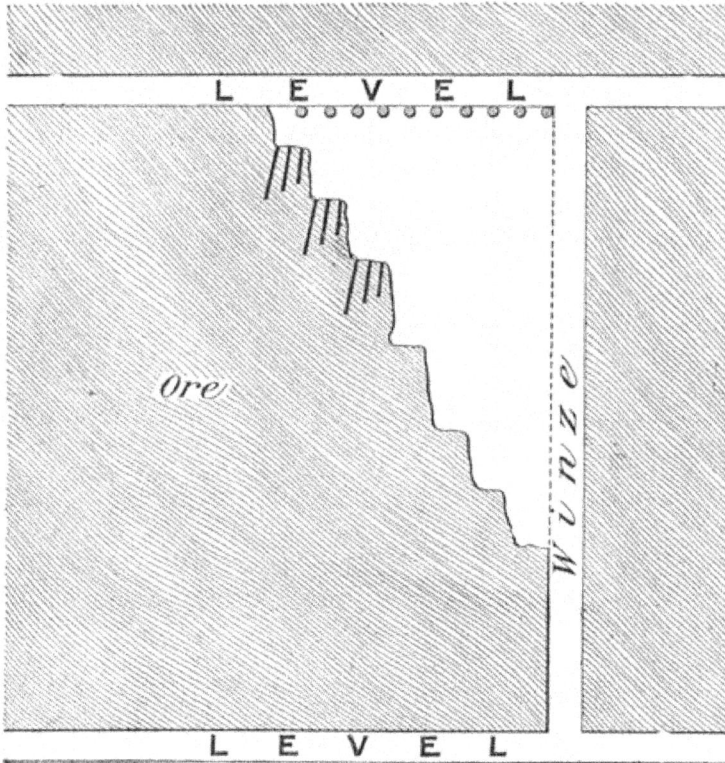

FIG. 22.

This system can be advantageously applied only in the rare cases in which the walls require little or no support, and where very little or no waste requiring separation is broken with the ore in the stopes. To support the walls in bad ground in underhand stopes would be far more costly than with overhand stopes, for square-set timbering would be most difficult to introduce, and to support the walls with waste and stulls would be even more troublesome. Any waste broken must needs be thrown up to the level above or be stored upon specially built stages—again a costly proceeding.

A further drawback lies in the fact that the broken ore

follows down the face of the stope, and must be shoveled off each bench. It thus all arrives at a single point,—the winze,— and must be drawn from a single ore-pass into the level. This usually results not only in more shoveling but in a congestion at the passes not present in overhand stoping, for with that method several chutes are available for discharging ore into the levels. Where the walls require no support and no selection is desired in the stopes, the advantage of the men standing on the solid ore to work, and of having all down holes and therefore drilled wet, gives this method a distinct place. In using this system, in order to protect the men, a pillar is often left under the level by driving a sublevel, the pillar being easily recoverable later. The method of sublevels is of advantage largely in avoiding the timbering of levels.

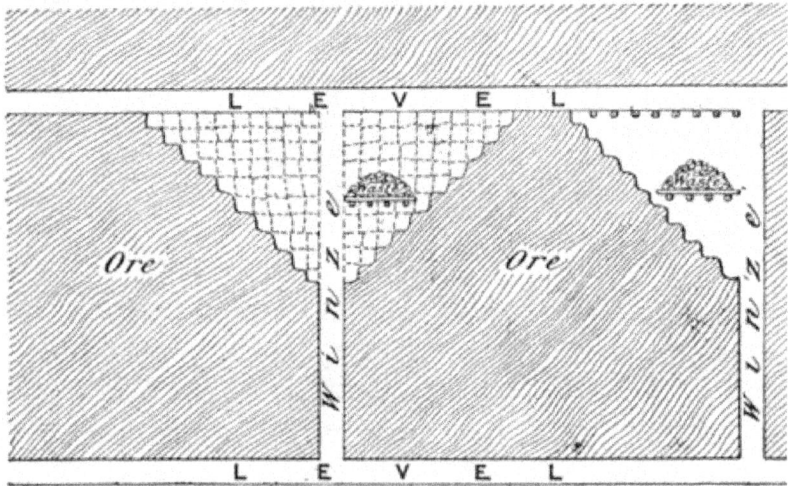

FIG.23.—Longitudinal section of an underhand stope.

Overhand Stopes.—By far the greatest bulk of ore is broken overhand, that is broken upward from one level to the next above. There are two general forms which such stopes are given,—"horizontal" and "rill."

The horizontal "flat-back" or "long-wall" stope, as it is variously called, shown in Figure 24, is operated by breaking the ore in slices parallel with the levels. In rill-stoping the ore is cut back from the winzes in such a way that a pyramid-shaped room is created, with its apex in the winze and its base at the level (Figs. 25 and 26). Horizontal or flat-backed stopes can be applied to almost any dip, while "rill-stoping" finds its

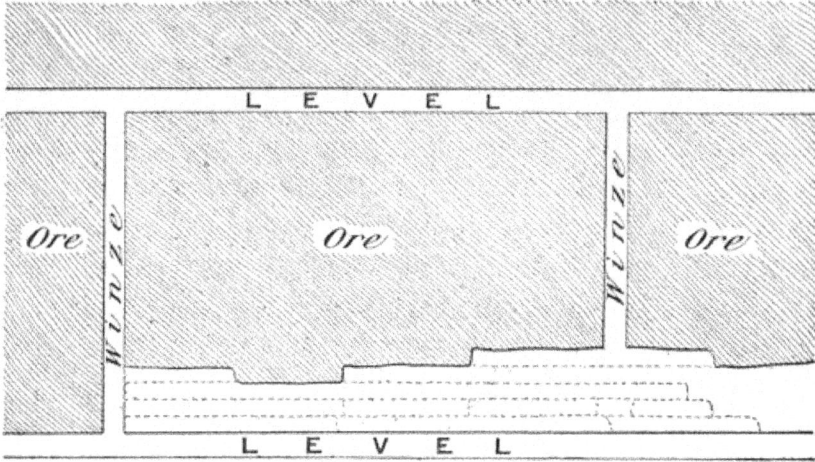

FIG.24.—Horizontal-cut overhand stope—longitudinal section.

most advantageous application where the dip is such that the ore will "run," or where it can be made to "run" with a little help. The particular application of the two systems is dependent not only on the dip but on the method of supporting the excavation and the ore. With rill-stoping, it is possible to

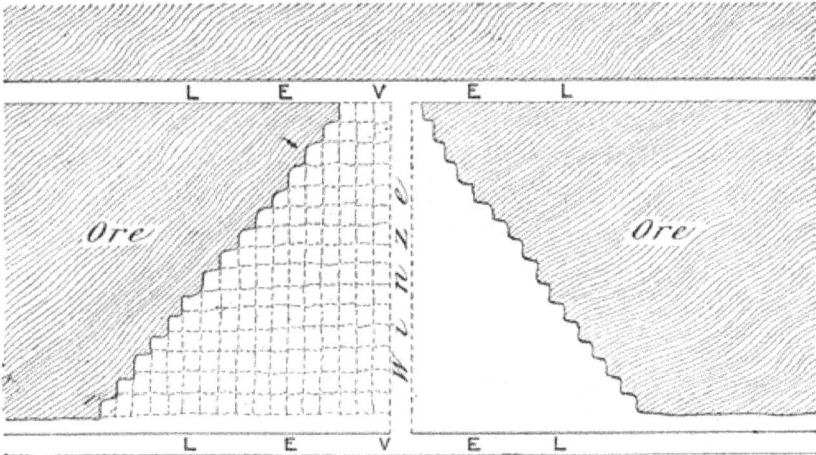

FIG.25.—Rill-cut overhand stope—longitudinal section.

cut the breaking benches back horizontally from the winzes (Fig. 25), or to stagger the cuts in such a manner as to take the slices in a descending angle (Figs. 21 and 26).

In the "rill" method of incline cuts, all the drill-holes are "down" holes (Fig. 21), and can be drilled wet, while in horizontal cuts or flat-backed stopes, at least part of the holes must be "uppers" (Fig. 20). Aside from the easier and cheaper drilling and setting up of machines with this kind of "cut," there is no drill dust,—a great desideratum in these days of miners' phthisis.

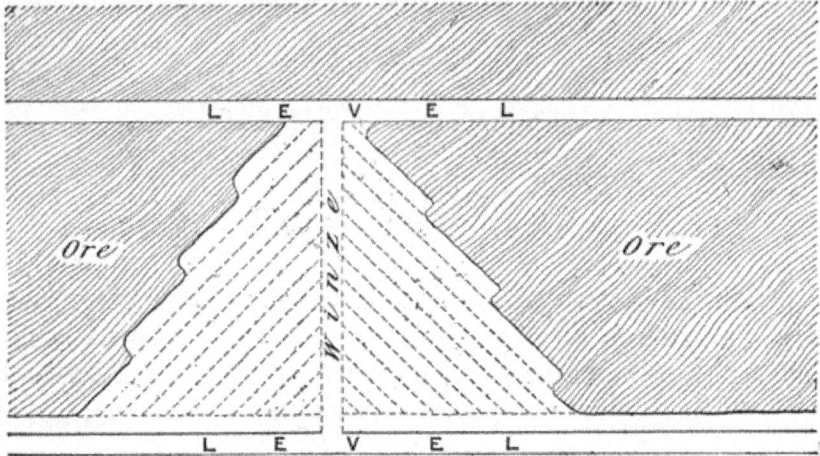

FIG.26.—Rill-cut overhand stope-longitudinal section.

A further advantage in the "rill" cut arises in cases where horizontal jointing planes run through the ore of a sort from which unduly large masses break away in "flat-back" stopes. By the descending cut of the "rill" method these calamities can be in a measure avoided. In cases of dips over 40° the greatest advantage in "rill" stoping arises from the possibility of pouring filling or timber into the stope from above with less handling, because the ore and material will run down the sides of the pyramid (Figs. 32 and 34). Thus not only is there less shoveling required, but fewer ore-passes and a less number of preliminary winzes are necessary, and a wider level interval is possible. This matter will be gone into more fully later.

Combined Stopes.—A combined stope is made by the coincident working of the underhand and "rill" method (Fig. 27). This order of stope has the same limitations in general as the underhand kind. For flat veins with strong walls, it has a great superiority in that the stope is carried back more or less parallel with the winzes, and thus broken ore after blasting lies in a line on the gradient of the stope. It is, therefore, conveniently placed for mechanical stope haulage. A further advantage is gained in that winzes may be placed long distances apart, and that men are not required, either when at work or passing to and from it, to be ever far from the face, and they are thus in the safest ground, so that timber and filling protection which may be otherwise necessary is not required. This method is largely used in South Africa.

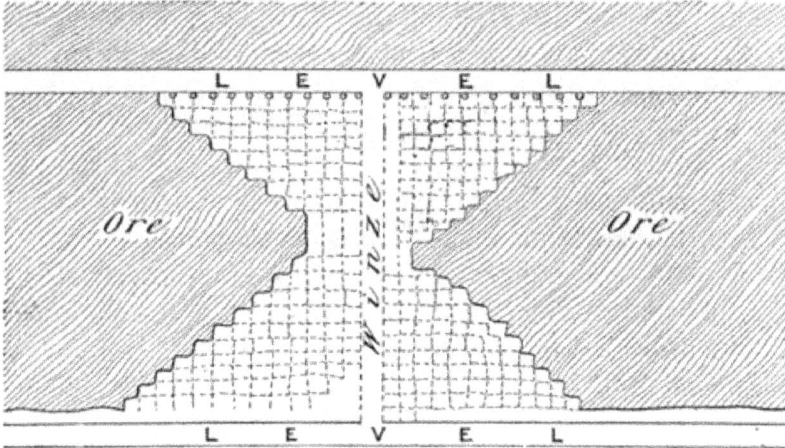

FIG.27.—Longitudinal section of a combined stope.

Minimum Width of Stopes.—The minimum stoping width which can be consistently broken with hand-holes is about 30 inches, and this only where there is considerable dip to the ore. This space is so narrow that it is of doubtful advantage in any case, and 40 inches is more common in narrow mines, especially where worked with white men. Where machine-drills are used about 4 feet is the minimum width feasible.

Resuing.—In very narrow veins where a certain amount of wall-rock must be broken to give working space, it pays under some circumstances to advance the stope into the wall-rock ahead of the ore, thus stripping the ore and enabling it to be broken separately. This permits of cleaner selection of the ore; but it is a problem to be worked out in each case, as to whether rough sorting of some waste in the stopes, or further sorting at surface with inevitable treatment of some waste rock, is more economical than separate stoping cuts and inevitably wider stopes.

Valuing Ore in Course of Breaking.—There are many ores whose payability can be determined by inspection, but there are many of which it cannot. Continuous assaying is in the latter cases absolutely necessary to avoid the treatment of valueless material. In such instances, sampling after each stoping-cut is essential, the unprofitable ore being broken down and used as waste. Where values fade into the walls, as in impregnation deposits, the width of stopes depends upon the limit of payability. In these cases, drill-holes are put into the walls and the drillings assayed. If the ore is found profitable, the holes are blasted out. The gauge of what is profitable in such situations is not dependent simply upon the average total working costs of the mine, for ore in that position can be said to cost nothing for development work and administration; moreover, it is usually more cheaply broken than the average breaking cost, men and machines being already on the spot.

CHAPTER XI.

METHODS OF SUPPORTING EXCAVATION.

TIMBERING; FILLING WITH WASTE; FILLING WITH
BROKEN ORE; PILLARS OF ORE; ARTIFICIAL
PILLARS; CAVING SYSTEM.

Most stopes require support to be given to the walls and often
to the ore itself. Where they do require support there are five
principal methods of accomplishing it. The application of any
particular method depends upon the dip, width of ore-body, character of the ore and walls, and cost of materials. The various
systems are by:—
1. Timbering.
2. Filling with waste.
3. Filling with broken ore subsequently withdrawn.
4. Pillars of ore.
5. Artificial pillars built of timbers and waste.
6. Caving.

Timbering.—At one time timbering was the almost universal
means of support in such excavations, but gradually various methods for the economical application of waste and ore itself have
come forward, until timbering is fast becoming a secondary device. Aside from economy in working without it, the dangers of
creeps, or crushing, and of fires are sufficient incentives to do
away with wood as far as possible.

There are three principal systems of timber support to excavations,—by stulls, square-sets, and cribs.

Stulls are serviceable only where the deposit is so narrow that
the opening can be bridged by single timbers between wall and
wall (Figs. 28 and 43). This system can be applied to any dip and
is most useful in narrow deposits where the walls are not too
heavy. Stulls in inclined deposits are usually set at a slightly
higher

angle than that perpendicular to the walls, in order that the vertical pressure of the hanging wall will serve to tighten them in

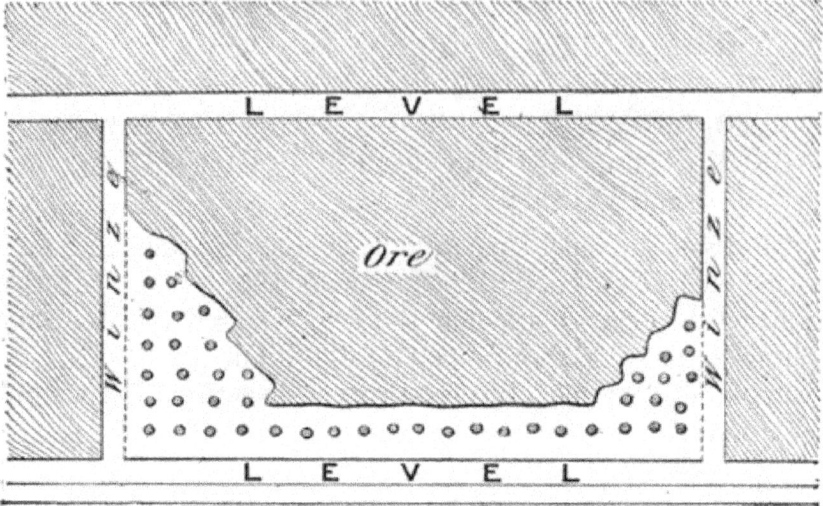

FIG.28.—Longitudinal section of stull-supported stope.

position. The "stull" system can, in inclined deposits, be further
strengthened by building waste pillars against them, in which
case the arrangement merges into the system of artificial pillars.

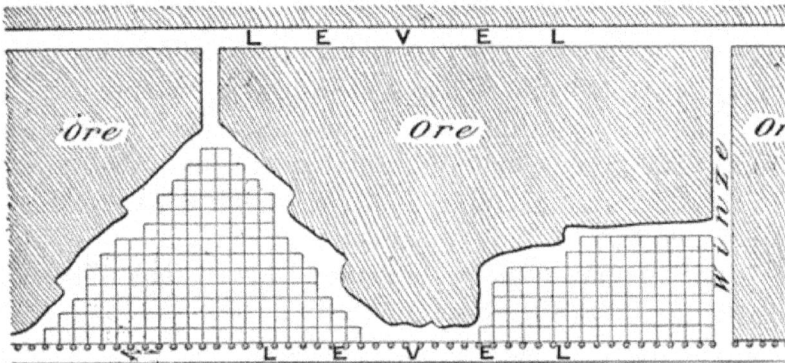

FIG.29.—Longitudinal section showing square-set timbering.

Square-sets (Figs. 29 and 30), that is, trusses built in the opening as
the ore is removed, are applicable to almost any dip or width of ore,
but generally are applied only in deposits too wide, or to rock too
heavy, for stulls. Such trusses are usually constructed on

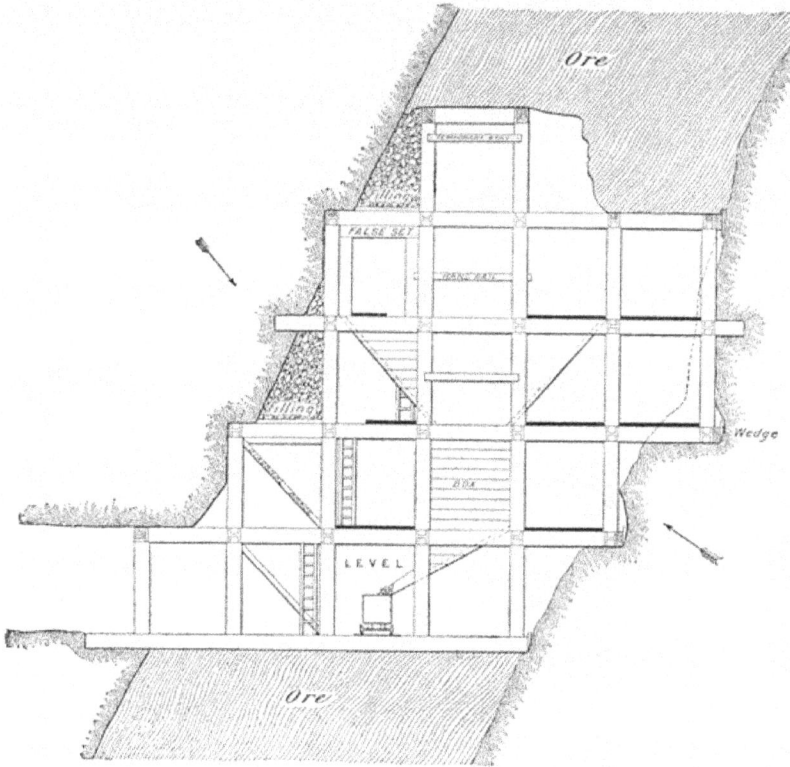

FIG.30.—Square-set timbering on inclined ore-body. Showing ultimate
strain on timbers.

vertical and horizontal lines, and while during actual ore-
breaking the strains are partially vertical, ultimately, however,
when the weight of the walls begins to be felt, these strains, except in vertical deposits, come at an angle to lines of strength in
the trusses, and therefore timber constructions of this type
present little ultimate resistance (Fig. 30). Square-set timbers are
sometimes set to present the maximum resistance to the direction
of strain, but the difficulties of placing them in position and variations in the direction of strain on various parts of the stope do
not often commend the method. As a general rule square-sets on
horizontal lines answer well enough for the period of actual ore-
breaking. The crushing or creeps is usually some time later;

Page 106 Principles of Mining

and if the crushing may damage the whole mine, their use is fraught with danger. Re-enforcement by building in waste is often resorted to. When done fully, it is difficult to see the utility of the enclosed timber, for entire waste-filling would in most cases be cheaper and equally efficient.

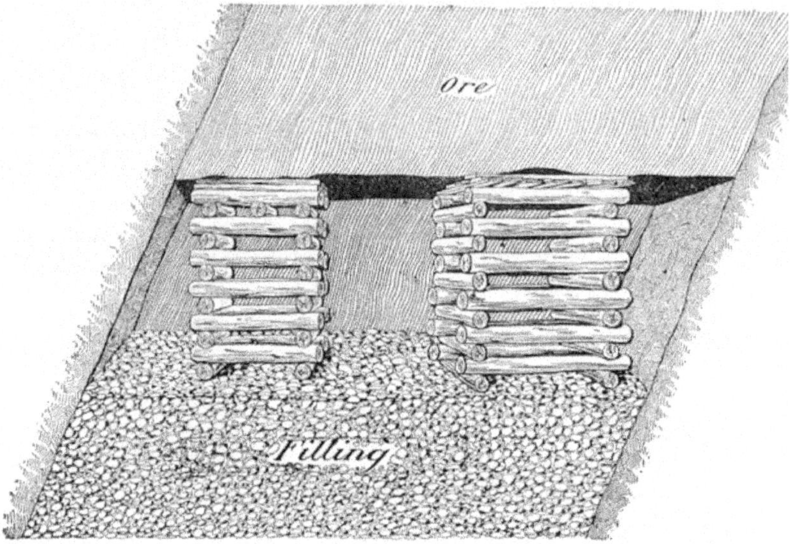

Fig.31.—"Cribs."

There is always, with wood constructions, as said before, the very pertinent danger of subsequent crushing and of subsidence in after years, and the great risk of fires. Both these disasters have cost Comstock and Broken Hill mines, directly or indirectly, millions of dollars, and the outlay on timber and repairs one way or another would have paid for the filling system ten times over. There are cases where, by virtue of the cheapness of timber, "square-setting" is the most economical method. Again, there are instances where the ore lies in such a manner—particularly in limestone replacements—as to preclude other means of support. These cases are being yearly more and more evaded by the ingenuity of engineers in charge. The author believes it soon will be recognized that the situation is rare indeed where complete square-setting is necessarily without an economical alternative. An objection is sometimes raised to filling in favor of timber, in that if it become desirable to restope the walls for low-grade ore left behind, such stopes could only be entered by drawing the filling, with consequent danger of total collapse. Such a contingency can be provided for in large ore-bodies by installing an outer shell of sets of timber around the periphery of the stope and filling the inside with waste. If the crushing possibilities are too great for this method then, the subsequent recovery of ore is hopeless in any event. In narrow ore-bodies with crushing walls recovery of ore once left behind is not often possible.

The third sort of timber constructions are cribs, a "log-house" sort of structure usually filled with waste, and more fully discussed under artificial pillars (Fig. 31). The further comparative merits of timbering with other methods will be analyzed as the different systems are described.

Filling with Waste.—The system of filling stope-excavations completely with waste in alternating progress with ore-breaking is of wide and increasingly general application (Figs. 32, 33, 34, 35).

Although a certain amount of waste is ordinarily available in the stopes themselves, or from development work in the mine, such a supply must usually be supplemented from other directions. Treatment residues afford the easiest and cheapest handled material. Quarried rock ranks next, and in default of any other easy supply, materials from crosscuts driven into the stope-walls are sometimes resorted to.

In working the system to the best advantage, the winzes through the block of ore under attack are kept in alignment with similar openings above, in order that filling may be poured through the mine from the surface or any intermediate point. Winzes to be used for filling should be put on the hanging-wall side of the area to be filled, for the filling poured down will then reach the foot-wall side of the stopes with a minimum of handling. In some instances, one special winze is arranged for passing all filling from the surface to a level above the principal stoping

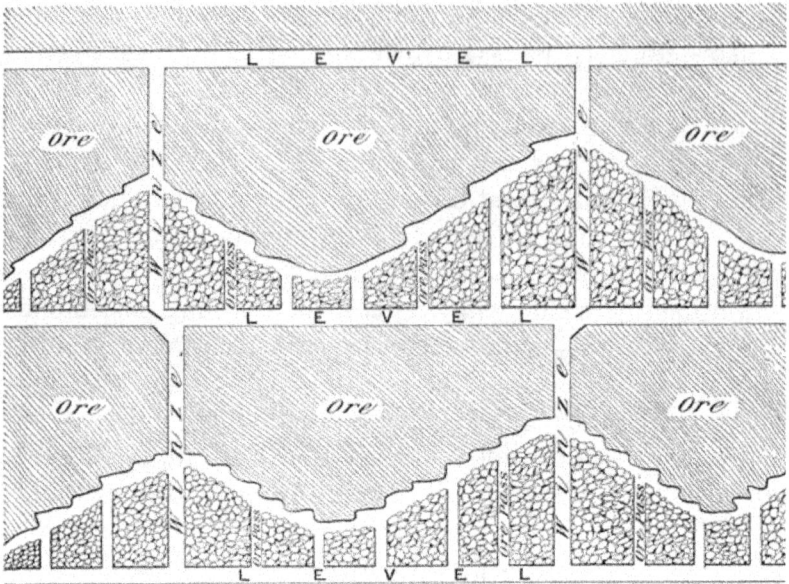

FIG.32.—Longitudinal section. Rill stope filled with waste.

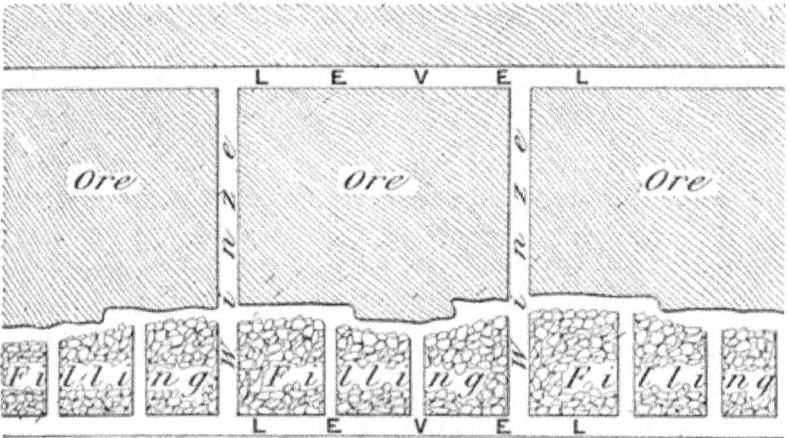

FIG.33.—Longitudinal section. Horizontal stope filled with waste.

operations; and it is then distributed along the levels into the winzes, and thus to the operating stopes, by belt-conveyors.

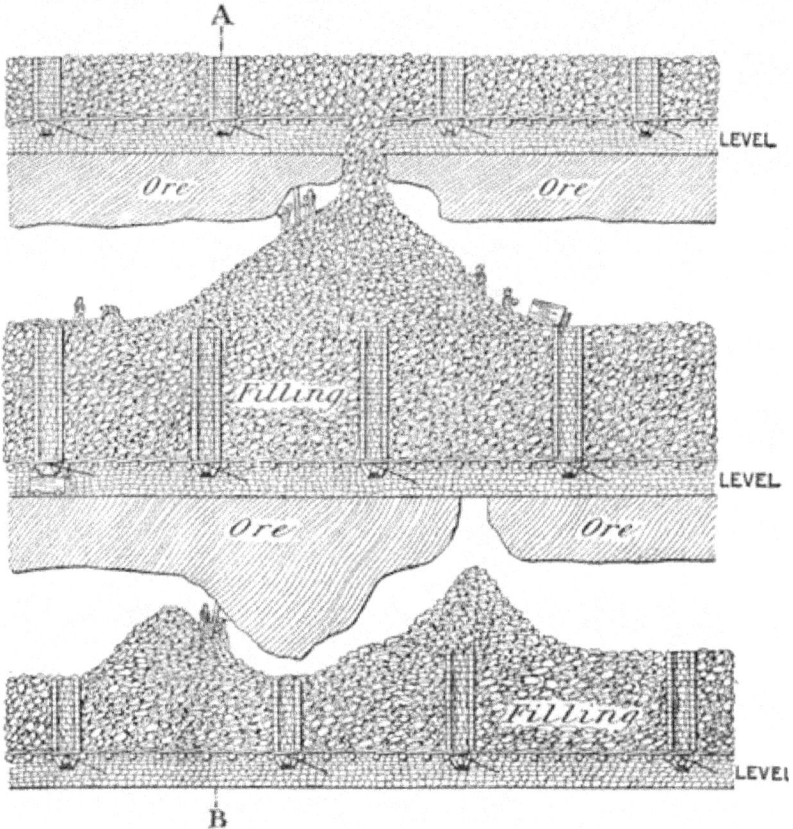

FIG.34.—Longitudinal section. Waste-filled stope with dry-walling of levels and passes.

In this system of stope support the ore is broken at intervals alternating with filling. If there is danger of much loss from mixing broken ore and filling, "sollars" of boards or poles are laid on the waste. If the ore is very rich, old canvas or cowhides are sometimes put under the boards. Before the filling interval, the ore passes are built close to the face above previous filling and their tops covered temporarily to prevent their being filled with running waste. If the walls are bad, the filling is kept close to the face. If the unbroken ore requires support, short stulls set on the waste (as in Fig. 39) are usually sufficient until the next cut is taken off, when the timber can be recovered. If stulls are insufficient, cribs or bulkheads (Fig. 31) are also used and often buried in the filling.

Both flat-backed and rill-stope methods of breaking are employed in conjunction with filled stopes. The advantages of the rill-stopes are so patent as to make it difficult to understand why they are not universally adopted when the dip permits their use at all. In rill-stopes (Figs. 32 and 34) the waste flows to its destination with a minimum of handling. Winzes and ore-passes are

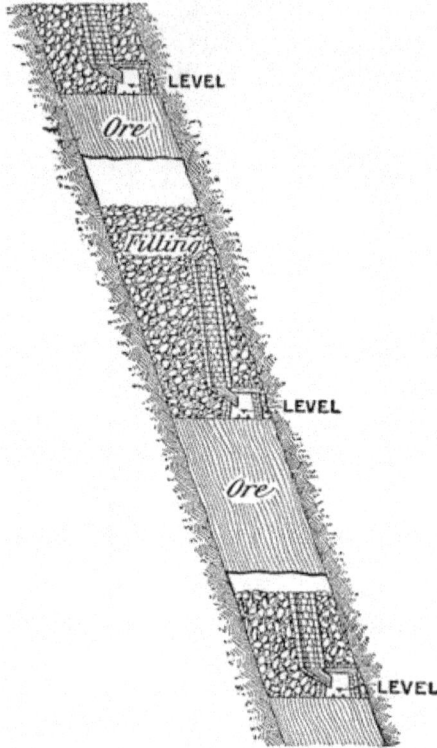

FIG.35.—Cross-section of Fig. 34 on line *A-B*.

not required with the same frequency as in horizontal breaking, and the broken ore always lies on the slope towards the passes and is therefore also easier to shovel. In flat-backed stopes (Fig. 33) winzes must be put in every 50 feet or so, while in rill-stopes they can be double this distance apart. The system is applicable by modification to almost any width of ore. It finds its most economical field where the dip of the stope floor is over 45°, when waste and ore, with the help of the "rill," will flow to their destination. For dips from under about 45° to about 30° or 35°, where the waste and ore will not "flow" easily, shoveling can be helped by the use of the "rill" system and often evaded altogether, if flow be assisted by a sheet-iron trough described in the discussion of stope transport. Further saving in shoveling can be gained in this method, by giving a steeper pitch to the filling winzes and to the ore-passes, by starting them from crosscuts in the wall, and by carrying them at greater angles than the pitch of the ore (Fig. 36). These artifices combined have worked out most economically on several mines within the writer's experience, with the dip as flat as 30°. For very flat dips, where filling is to be employed, rill-stoping has no advantage over flat-backed cuts, and in such cases it is often advisable to assist stope transport by temporary tracks and cars which obviously could not be worked on the tortuous contour of a rill-stope, so that for dips under 30° advantage lies with "flat-backed" ore-breaking.

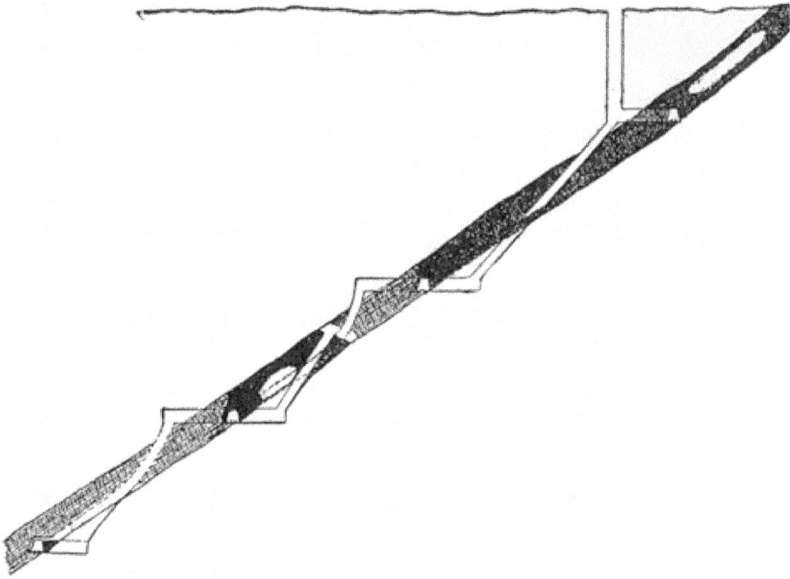

FIG. 36.—Cross-section showing method of steepening winzes and ore passes.

On very wide ore-bodies where the support of the standing ore itself becomes a great problem, the filling system can be applied by combining it with square-setting. In this case the stopes are carried in panels laid out transversally to the strike as wide as the standing strength of the ore permits. On both sides of each panel a fence of lagged square-sets is carried up and the area between is filled with waste. The panels are stoped out alternately. The application of this method at Broken Hill will be described later. (See pages 120 and Figs. 41 and 42.) The same type of wide ore-body can be managed also on the filling system by the use of frequent "bulkheads" to support the ore (Fig. 31).

Compared with timbering methods, filling has the great advantage of more effective support to the mine, less danger of creeps, and absolute freedom from the peril of fire. The relative expense of the two systems is determined by the cost of materials and labor. Two extreme cases illustrate the result of these economic factors with sufficient clearness. It is stated that the cost of timbering stopes on the Le Roi Mine by square-sets is about 21 cents per ton of ore excavated. In the Ivanhoe mine of West Australia the cost of filling stopes with tailings is about 22 cents per ton of ore excavated. At the former mine the average cost of timber is under $10 per M board-measure, while at the latter its price would be $50 per M board-measure; although labor is about of the same efficiency and wage, the cost in the Ivanhoe by square-setting would be about 65 cents per ton of ore broken. In the Le Roi, on the other hand, no residues are available for filling. To quarry rock or drive crosscuts into the walls might make this system cost 65 cents per ton of ore broken if applied to that mine. The comparative value of the filling method with other systems will be discussed later.

Filling with Broken Ore subsequently Withdrawn.—This order of support is called by various names, the favorite being "shrinkage-stoping." The method is to break the ore on to the roof of the level, and by thus filling the stope with broken ore, provide temporary support to the walls and furnish standing floor upon which to work in making the next cut (Figs. 37, 38, and 39.) As broken material occupies 30 to 40% more space than rock *in situ*, in order to provide working space at the face, the broken ore must be drawn from along the level after each cut. When the area attacked is completely broken through from level to level, the stope will be full of loose broken ore, which is then entirely drawn off.

A block to be attacked by this method requires preliminary winzes only at the extremities of the stope,—for entry and for ventilation. Where it is desired to maintain the winzes after stoping, they must either be strongly timbered and lagged on the stope side, be driven in the walls, or be protected by a pillar of ore (Fig. 37). The settling ore and the crushing after the stope is empty make it difficult to maintain timbered winzes.

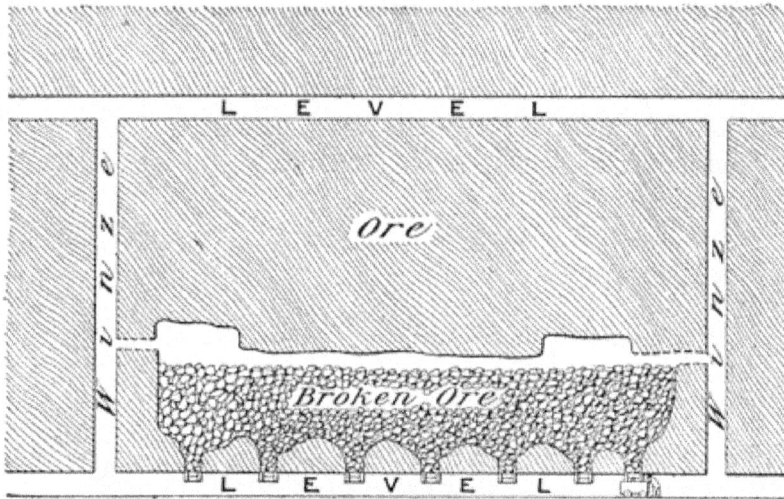

FIG.37.—Longitudinal section of stope filled with broken ore.

Where it can be done without danger to the mine, the empty stopes are allowed to cave. If such crushing would be dangerous, either the walls must be held up by pillars of unbroken ore, as in the Alaska Treadwell, where large "rib" pillars are left, or the open spaces must be filled with waste. Filling the empty stope is usually done by opening frequent passes along the base of the filled stope above, and allowing the material of the upper stope to flood the lower one. This program continued upwards through the mine allows the whole filling of the mine to descend gradually and thus requires replenishment only into the top. The old stopes in the less critical and usually exhausted territory nearer the surface are sometimes left without replenishing their filling.

The weight of broken ore standing at such a high angle as to settle rapidly is very considerable upon the level; moreover, at the moment when the stope is entirely drawn off, the pressure of the walls as well is likely to be very great. The roadways in this system therefore require more than usual protection. Three methods are used: (*a*) timbering; (*b*) driving a sublevel in the ore above the main roadway as a stoping-base, thus leaving a pillar of ore over the roadway (Fig. 39); (*c*) by dry-walling the levels, as in the Baltic mine, Michigan (Figs. 34 and 35). By the use of sublevels the main roadways are sometimes driven in the walls (Fig. 38) and in many cases all timbering is saved. To recover pillars left below sublevels is a rather difficult task, especially if the old stope above is caved or filled. The use of pillars in substitution for timber, if the pillars are to be lost, is simply a matter of economics as to whether the lost ore would repay the cost of other devices.

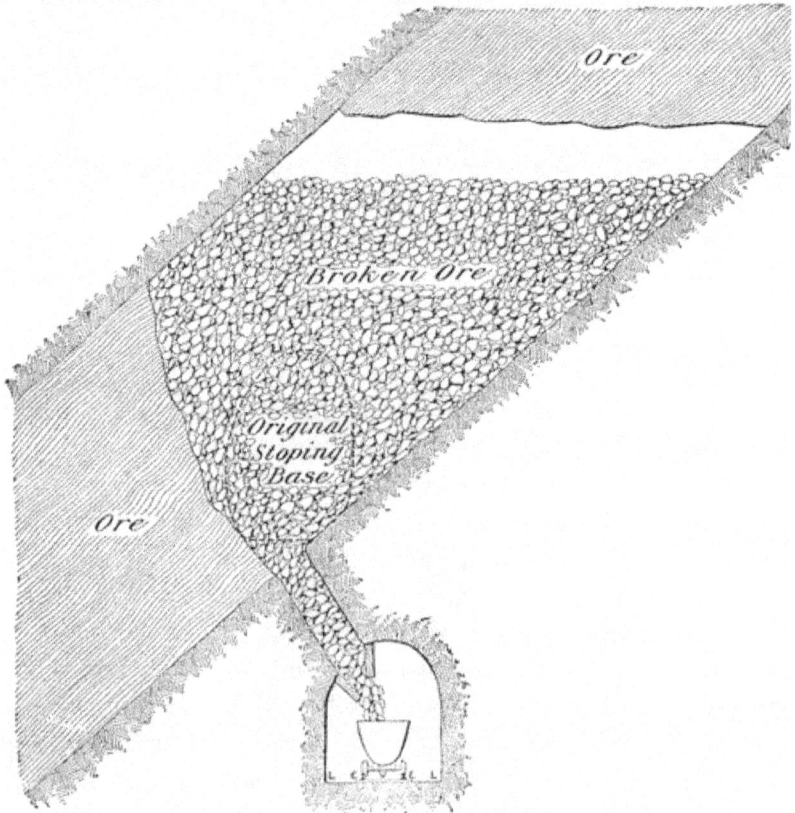

FIG.38.—Cross-section of "shrinkage" stope.

Frequent ore-chutes through the level timbers, or from the sublevels, are necessary to prevent lodgment of broken ore be- tween such passes, because it is usually too dangerous for men to enter the emptying stope to shovel out the lodged remnants. Where the ore-body is wide, and in order that there may be no lodgment of ore, the timbers over the level are set so as to form a trough along the level; or where pillars are left, they are made "A"-shaped between the chutes, as indicated in Figure 37.

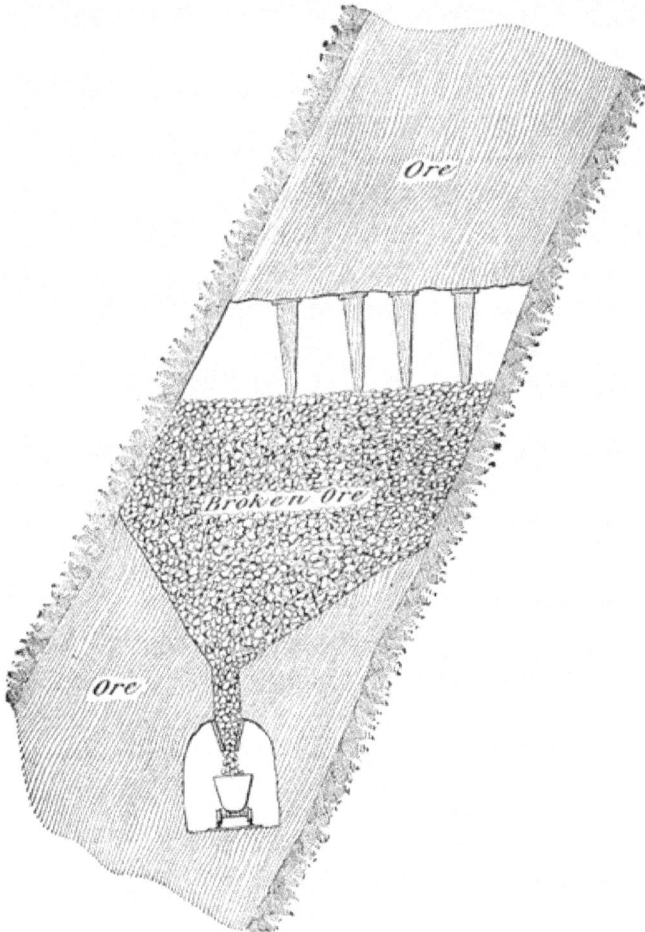

FIG.39.—Cross-section of "shrinkage" stope.

The method of breaking the ore in conjunction with this means of support in comparatively narrow deposits can be on the rill, in order to have the advantage of down holes. Usually, how- ever, flat-back or horizontal cuts are desirable, as in such an arrangement it is less troublesome to regulate the drawing of the ore so as to provide proper head room. Where stopes are wide, ore is sometimes cut arch-shaped from wall to wall to assure its standing. Where this method of support is not of avail, short, sharply tapering stulls are put in from the broken ore to the face (Fig. 39). When the cut above these stulls is taken out, they are pulled up and are used again.

This method of stoping is only applicable when:—

1. The deposit dips over 60°, and thus broken material will freely settle downward to be drawn off from the bottom.

2. The ore is consistently payable in character. No selection can be done in breaking, as all material broken must be drawn off together.

3. The hanging wall is strong, and will not crush or spall off waste into the ore.

4. The ore-body is regular in size, else loose ore will lodge on the foot wall. Stopes opened in this manner when partially empty are too dangerous for men to enter for shoveling out remnants.

The advantages of this system over others, where it is applicable, are:—

(a) A greater distance between levels can be operated and few winzes and rises are necessary, thus a great saving of development work can be effected. A stope 800 to 1000 feet long can be operated with a winze at either end and with levels 200 or 220 feet apart.

(b) There is no shoveling in the stopes at all.

(c) No timber is required. As compared with timbering by stulling, it will apply to stopes too wide and walls too heavy for this method. Moreover, little staging is required for working the face, since ore can be drawn from below in such a manner as to allow just the right head room.

(d) Compared to the system of filling with waste, coincidentally with breaking (second method), it saves altogether in some cases the cost of filling. In any event, it saves the cost of ore-passes, of shoveling into them, and of the detailed distribution of the filling.

Compared with other methods, the system has the following disadvantages, that:

A. The ore requires to be broken in the stopes to a degree of fineness which will prevent blocking of the chutes at the level. When pieces too large reach the chutes, nothing will open them but blasting,—to the damage of timbers and chutes. Some large rocks are always liable to be buried in the course of ore-breaking.

B. Practically no such perfection of walls exists, but some spalling of waste into the ore will take place. A crushing of the walls would soon mean the loss of large amounts of ore.

C. There is no possibility of regulating the mixture of grade of ore by varying the working points. It is months after the ore is broken before it can reach the levels.

D. The breaking of 60% more ore than immediate treatment demands results in the investment of a considerable sum of money. An equilibrium is ultimately established in a mine worked on this system when a certain number of stopes full of completely broken ore are available for entire withdrawal, and there is no further accumulation. But, in any event, a considerable amount of broken ore must be held in reserve. In one mine worked on this plan, with which the writer has had experience, the annual production is about 250,000 tons and the broken ore represents an investment which, at 5%, means an annual loss of interest amounting to 7 cents per ton of ore treated.

E. A mine once started on the system is most difficult to alter, owing to the lack of frequent winzes or passes. Especially is this so if the only alternative is filling, for an alteration to the system of filling coincident with breaking finds the mine short of filling winzes. As the conditions of walls and ore often alter with depth, change of system may be necessary and the situation may become very embarrassing.

F. The restoping of the walls for lower-grade ore at a later period is impossible, for the walls of the stope will be crushed, or, if filled with waste, will usually crush when it is drawn off to send to a lower stope.

The system has much to recommend it where conditions

are favorable. Like all other alternative methods of mining, it requires the most careful study in the light of the special conditions involved. In many mines it can be used for some stopes where not adaptable generally. It often solves the problem of blind ore-bodies, for they can by this means be frequently worked with an opening underneath only. Thus the cost of driving a roadway overhead is avoided, which would be required if timber or coincident filling were the alternatives. In such cases ventilation can be managed without an opening above, by so directing the current of air that it will rise through a winze from the level below, flow along the stope and into the level again at the further end of the stope through another winze.

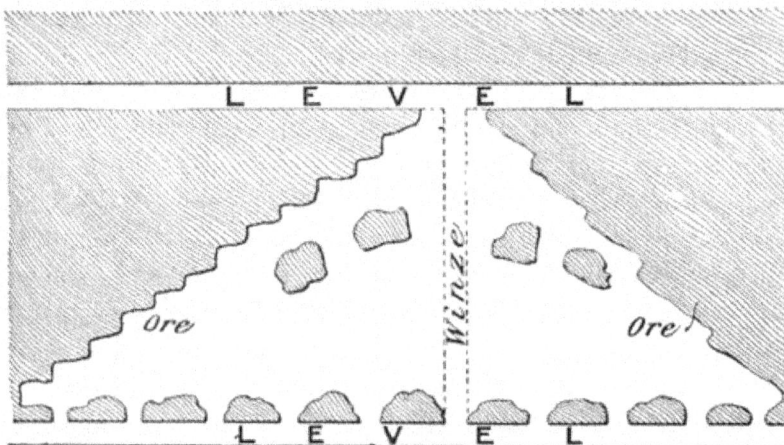

FIG.40.—Longitudinal section. Ore-pillar support in narrow stopes.

Support by Pillars of Ore.—As a method of mining metals of the sort under discussion, the use of ore-pillars except in conjunction with some other means of support has no general application. To use them without assistance implies walls sufficiently strong to hold between pillars; to leave them permanently anywhere implies that the ore abandoned would not repay the labor and the material of a substitute. There are cases of large, very low-grade mines where to abandon one-half the ore as pillars is more profitable than total extraction, but the margin of payability in such ore must be very, very narrow. Unpayable spots are always left as pillars, for obvious reasons.

Permanent ore-pillars as an adjunct to other methods of support are in use. Such are the rib-pillars in the Alaska Treadwell,
the form of which is indicated by the upward extension of the
pillars adjacent to the winzes, shown in Figure 37. Always a
careful balance must be cast as to the value of the ore left, and as
to the cost of a substitute, because every ore-pillar can be removed at some outlay. Temporary pillars are not unusual,
particularly to protect roadways and shafts. They are, when left
for these purposes, removed ultimately, usually by beginning at
the farther end and working back to the final exit.

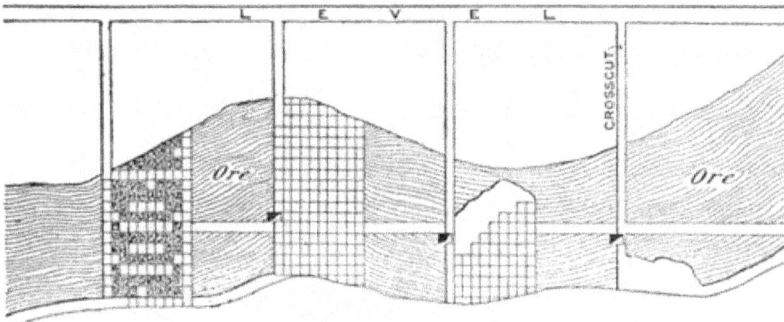

FIG.41.—Horizontal plan at levels of Broken Hill. Method of alternate
stopes and ore-pillars.

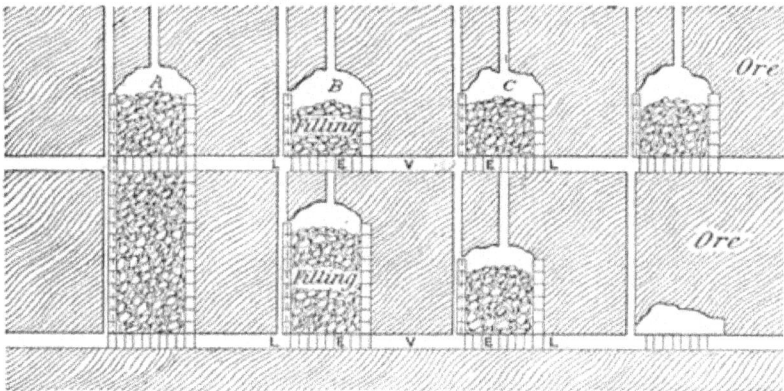

FIG.42.—Longitudinal section of Figure 41.

A form of temporary ore-pillars in very wide deposits is made
use of in conjunction with both filling and timbering (Figs. 37,
39, 40). In the use of temporary pillars for ore-bodies

100 to 250 feet wide at Broken Hill, stopes are carried up at right angles to the strike, each fifty feet wide and clear across the ore-body (Figs. 41 and 42). A solid pillar of the same width is left in the first instance between adjacent stopes, and the initial series of stopes are walled with one square-set on the sides as the stope is broken upward. The room between these two lines of sets is filled with waste alternating with ore-breaking in the usual filling method. When the ore from the first group of alternate stopes (*ABC*, Fig. 42) is completely removed, the pillars are stoped out and replaced with waste. The square-sets of the first set of stopes thus become the boundaries of the second set. Entry and ventilation are obtained through these

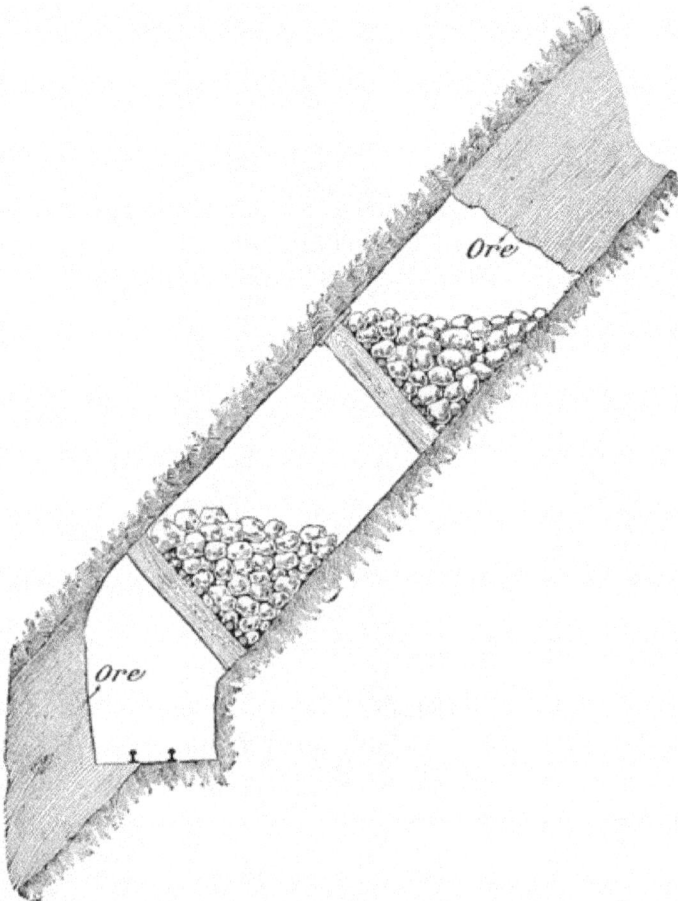

FIG.43.—Cross-section of stull support with waste re-enforcement.

lines of square-sets, and the ore is passed out of the stopes through them.

Artificial Pillars.—This system also implies a roof so strong as not to demand continuous support. Artificial pillars are built in many different ways. The method most current in fairly narrow deposits is to re-enforce stulls by packing waste above them (Figs. 43 and 44). Not only is it thus possible to economize in stulls by using the waste which accumulates underground, but the principle applies also to cases where the stulls alone are not sufficient support, and yet where complete filling or square-setting is unnecessary. When the conditions are propitious for this method, it has the comparative advantage over timber systems of saving timber, and over filling systems of saving imported filling. Moreover, these constructions being pillar-shaped (Fig. 44), the intervals between them provide outlets for broken ore, and specially built passes are unnecessary. The method has two disadvantages as against the square-set or filling process, in that more staging must be provided from which to work, and in stopes over six feet the erection of machine-drill columns is tedious and costly in time and wages.

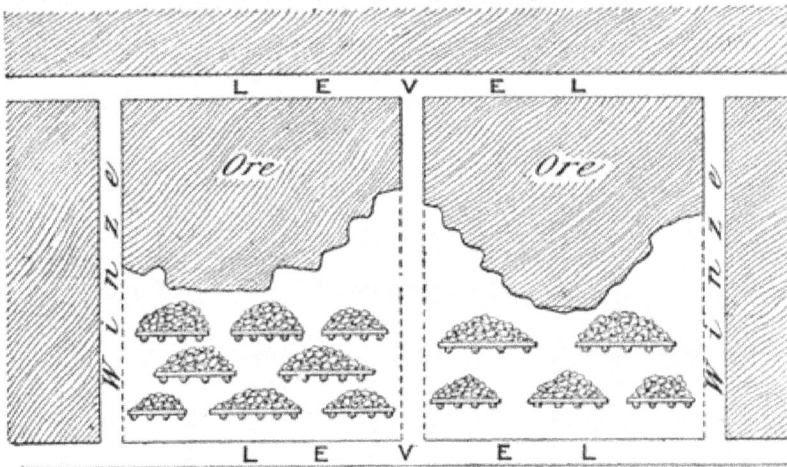

FIG.44.—Longitudinal section of stull and waste pillars.

In wide deposits of markedly flat, irregular ore-bodies, where a definite system is difficult and where timber is expensive, cribs of cord-wood or logs filled with waste after the order shown in

Figure 31, often make fairly sound pillars. They will not last in-definitely and are best adapted to the temporary support of the ore-roof pending filling. The increased difficulty in setting up machine drills in such stopes adds to the breaking costs,—often enough to warrant another method of support.

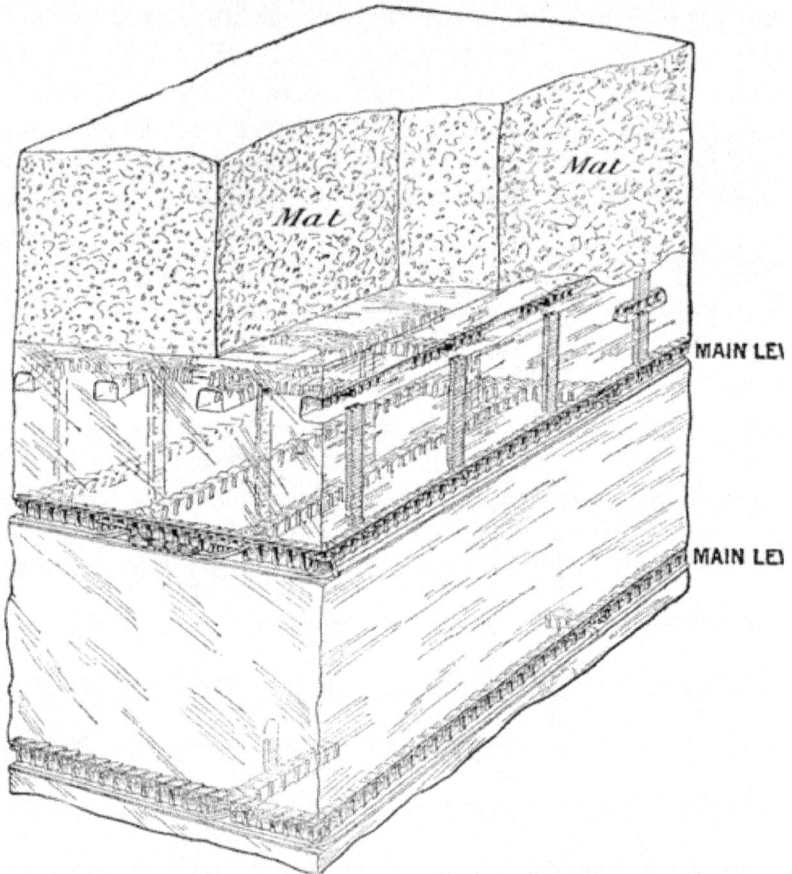

FIG.45.—Sublevel caving system.

Caving Systems.—This method, with variations, has been applied to large iron deposits, to the Kimberley diamond mines, to some copper mines, but in general it has little application to the metal mines under consideration, as few ore-bodies are of sufficiently large horizontal area. The system is dependent upon a large area of loose or "heavy" ground pressing directly on the ore with weight, such that if the ore be cut into pillars, these will crush. The details of the system vary, but in general the *modus operandi* is to prepare roadways through the ore, and from the roadways to put rises, from which sublevels are driven close under the floating mass of waste and ore,—sometimes called the "matte" (Fig. 45). The pillars between these sublevels are then cut away until the weight above crushes them down. When all the crushed ore which can be safely reached is extracted, retreat is made and another series of subopenings is then driven close under the "matte." The pillar is reduced until it crushes and the operation is repeated. Eventually the bottom strata of the "matte" become largely ore, and a sort of equilibrium is reached when there is not much loss in this direction. "Top slicing" is a variation of the above method by carrying a horizontal stope from the rises immediately under the matte, supporting the floating material with timber. At Kimberley the system is varied in that galleries are run out to the edge of the diamond-iferous area and enlarged until the pillar between crushes.

In the caving methods, between 40 and 50% of the ore is removed by the preliminary openings, and as they are all headings of some sort, the average cost per ton of this particular ore is higher than by ordinary stoping methods. On the other hand, the remaining 50 to 60% of the ore costs nothing to break, and the average cost is often remarkably low. As said, the system implies bodies of large horizontal area. They must start near enough to the surface that the whole superincumbent mass may cave and give crushing weight, or the immediately overhanging roof must easily cave. All of these are conditions not often met with in mines of the character under review.

CHAPTER XII.

MECHANICAL EQUIPMENT.

CONDITIONS BEARING ON MINE EQUIPMENT;
WINDING APPLIANCES; HAULAGE EQUIPMENT IN
SHAFTS; LATERAL UNDERGROUND TRANSPORT;
TRANSPORT IN STOPES.

There is no type of mechanical engineering which presents such complexities in determination of the best equipment as does that of mining. Not only does the economic side dominate over pure mechanics, but machines must be installed and operated under difficulties which arise from the most exceptional and conflicting conditions, none of which can be entirely satisfied. Compromise between capital outlay, operating efficiency, and conflicting demands is the key-note of the work.

These compromises are brought about by influences which lie outside the questions of mechanics of individual machines, and are mainly as follows:—

1. Continuous change in horizon of operations.
2. Uncertain life of the enterprise.
3. Care and preservation of human life.
4. Unequal adaptability of power transmission mediums.
5. Origin of power.

First.—The depth to be served and the volume of ore and water to be handled, are not only unknown at the initial equipment, but they are bound to change continuously in quantity, location, and horizon with the extension of the workings.

Second.—From the mine manager's point of view, which must embrace that of the mechanical engineer, further difficulty presents itself because the life of the enterprise is usually unknown, and therefore a manifest necessity arises for an economic balance of capital outlay and of operating efficiency commensurate with the prospects of the mine. Moreover, the initial capital is often limited, and makeshifts for this reason alone must be provided. In net result, no mineral deposit of speculative ultimate volume of ore warrants an initial equipment of the sort that will meet every eventuality, or of the kind that will give even the maximum efficiency which a free choice of mining machinery could obtain.

Third.—In the design and selection of mining machines, the safety of human life, the preservation of the health of workmen under conditions of limited space and ventilation, together with reliability and convenience in installing and working large mechanical tools, all dominate mechanical efficiency. For example, compressed-air transmission of power best meets the requirements of drilling, yet the mechanical losses in the generation, the transmission, and the application of compressed air probably total, from first to last, 70 to 85%.

Fourth.—All machines, except those for shaft haulage, must be operated by power transmitted from the surface, as obviously power generation underground is impossible. The conversion of power into a transmission medium and its transmission are, at the outset, bound to be the occasions of loss. Not only are the various forms of transmission by steam, electricity, compressed air, or rods, of different efficiency, but no one system lends itself to universal or economical application to all kinds of mining machines. Therefore it is not uncommon to find three or four different media of power transmission employed on the same mine. To illustrate: from the point of view of safety, reliability, control, and in most cases economy as well, we may say that direct steam is the best motive force for winding-engines; that for mechanical efficiency and reliability, rods constitute the best media of power transmission to pumps; that, considering ventilation and convenience, compressed air affords the best medium for drills. Yet there are other conditions as to character of the work, volume of water or ore, and the origin of power which must in special instances modify each and every one of these generalizations. For example, although pumping water with compressed air is mechanically the most inefficient of

devices, it often becomes the most advantageous, because compressed air may be of necessity laid on for other purposes, and
the extra power required to operate a small pump may be thus
most cheaply provided.

Fifth.—Further limitations and modifications arise out of the
origin of power, for the sources of power have an intimate bearing on the type of machine and media of transmission. This very
circumstance often compels giving away efficiency and convenience in some machines to gain more in others. This is evident
enough if the principal origins of power generation be examined.
They are in the main as follows:—

   a. Water-power available at the mine.

   b. Water-power available at a less distance than three
   or four miles.

   c. Water-power available some miles away, thus necessitating
   electrical transmission (or purchased electrical power).

   d. Steam-power to be generated at the mine.

   e. Gas-power to be generated at the mine.

   a. With water-power at the mine, winding engines can be operated by direct hydraulic application with a gain in economy
over direct steam, although with the sacrifice of control and reliability. Rods for pumps can be driven directly with water, but
this superiority in working economy means, as discussed later, a
loss of flexibility and increased total outlay over other forms of
transmission to pumps. As compressed air must be transmitted
for drills, the compressor would be operated direct from water-
wheels, but with less control in regularity of pressure delivery.

   b. With water-power a short distance from the mine, it would
normally be transmitted either by compressed air or by electricity. Compressed-air transmission would better satisfy winding
and drilling requirements, but would show a great comparative
loss in efficiency over electricity when applied to pumping. Despite the latter drawback, air transmission is a method growing
in favor, especially in view of the advance made in effecting
compression by falling water.

c. In the situation of transmission too far for using compressed air, there is no alternative but electricity. In these cases, direct electric winding is done, but under such disadvantages that it requires a comparatively very cheap power to take precedence over a subsidiary steam plant for this purpose. Electric air-compressors work under the material disadvantage of constant speed on a variable load, but this installation is also a question of economics. The pumping service is well performed by direct electrical pumps.

d. In this instance, winding and air-compression are well accomplished by direct steam applications; but pumping is beset with wholly undesirable alternatives, among which it is difficult to choose.

e. With internal combustion engines, gasoline (petrol) motors have more of a position in experimental than in systematic mining, for their application to winding and pumping and drilling is fraught with many losses. The engine must be under constant motion, and that, too, with variable loads. Where power from producer gas is used, there is a greater possibility of installing large equipments, and it is generally applied to the winding and lesser units by conversion into compressed air or electricity as an intermediate stage.

One thing becomes certain from these examples cited, that the right installation for any particular portion of the mine's equipment cannot be determined without reference to all the others. The whole system of power generation for surface work, as well as the transmission underground, must be formulated with regard to furnishing the best total result from all the complicated primary and secondary motors, even at the sacrifice of some members.

Each mine is a unique problem, and while it would be easy to sketch an ideal plant, there is no mine within the writer's knowledge upon which the ideal would, under the many variable conditions, be the most economical of installation or the most efficient of operation. The dominant feature of the task is an endeavor to find a compromise between efficiency and capital outlay. The result is a series of choices between unsatisfying alternatives, a number of which are usually found to have been wrong upon further extension of the mine in depth.

In a general way, it may be stated that where power is generated on the mine, economy in labor of handling fuel, driving engines, generation and condensing steam where steam is used, demand a consolidated power plant for the whole mine equipment. The principal motors should be driven direct by steam or gas, with power distribution by electricity to all outlying surface motors and sometimes to underground motors, and also to some underground motors by compressed air.

Much progress has been made in the past few years in the perfection of larger mining tools. Inherently many of our devices are of a wasteful character, not only on account of the need of special forms of transmission, but because they are required to operate under greatly varying loads. As an outcome of transmission losses and of providing capacity to cope with heavy peak loads, their efficiency on the basis of actual foot-pounds of work accomplished is very low.

The adoption of electric transmission in mine work, while in certain phases beneficial, has not decreased the perplexity which arises from many added alternatives, none of which are as yet a complete or desirable answer to any mine problem. When a satisfactory electric drill is invented, and a method is evolved of applying electricity to winding-engines that will not involve such abnormal losses due to high peak load then we will have a solution to our most difficult mechanical problems, and electricity will deserve the universal blessing which it has received in other branches of mechanical engineering.

It is not intended to discuss mine equipment problems from the machinery standpoint,—there are thousands of different devices,—but from the point of view of the mine administrator who finds in the manufactory the various machines which are applicable, and whose work then becomes that of choosing, arranging, and operating these tools.

The principal mechanical questions of a mine may be examined under the following heads:—

1. Shaft haulage.
2. Lateral underground transport.
3. Drainage.
4. Rock drilling.
5. Workshops.
6. Improvements in equipment.

SHAFT HAULAGE.

Winding Appliances.—No device has yet been found to displace the single load pulled up the shaft by winding a rope on a drum. Of driving mechanisms for drum motors the alternatives are the steam-engine, the electrical motor, and infrequently water-power or gas engines.

All these have to cope with one condition which, on the basis of work accomplished, gives them a very low mechanical efficiency. This difficulty is that the load is intermittent, and it must be started and accelerated at the point of maximum weight, and from that moment the power required diminishes to less than nothing at the end of the haul. A large number of devices are in use to equalize partially the inequalities of the load at different stages of the lift. The main lines of progress in this direction have been:—

a. The handling of two cages or skips with one engine or motor, the descending skip partially balancing the ascending one.
b. The use of tail-ropes or balance weights to compensate the increasing weight of the descending rope.
c. The use of skips instead of cages, thus permitting of a greater percentage of paying load.
d. The direct coupling of the motor to the drum shaft.
e. The cone-shaped construction of drums,—this latter being now largely displaced by the use of the tail-rope.

The first and third of these are absolutely essential for anything like economy and speed; the others are refinements depending on the work to be accomplished and the capital available.

Steam winding-engines require large cylinders to start the load, but when once started the requisite power is much reduced and the load is too small for steam economy. The throttling of the engine for controlling speed and reversing the engine at periodic stoppages militates against the maximum expansion and condensation of the steam and further increases the steam consumption. In result, the best of direct compound condensing engines consume from 60 to 100 pounds of steam per horse-power hour, against a possible efficiency of such an engine working under constant load of less than 16 pounds of steam per horse-power hour.

It is only within very recent years that electrical motors have been applied to winding. Even yet, all things considered, this application is of doubtful value except in localities of extremely cheap electrical power. The constant speed of alternating current motors at once places them at a disadvantage for this work of high peak and intermittent loads. While continuous-current motors can be made to partially overcome this drawback, such a current, where power is purchased or transmitted a long distance, is available only by conversion, which further increases the losses. However, schemes of electrical winding are in course of development which bid fair, by a sort of storage of power in heavy fly-wheels or storage batteries after the peak load, to reduce the total power consumption; but the very high first cost so far prevents their very general adoption for metal mining.

Winding-engines driven by direct water- or gas-power are of too rare application to warrant much discussion. Gasoline driven hoists have a distinct place in prospecting and early-stage mining, especially in desert countries where transport and fuel conditions are onerous, for both the machines and their fuel are easy of transport. As direct gas-engines entail constant motion of the engine at the power demand of the peak load, they are hopeless in mechanical efficiency.

Like all other motors in mining, the size and arrangement

of the motor and drum are dependent upon the duty which they will be called upon to perform. This is primarily dependent upon the depth to be hoisted from, the volume of the ore, and the size of the load. For shallow depths and tonnages up to, say, 200 tons daily, geared engines have a place on account of their low capital cost. Where great rope speed is not essential they are fully as economical as direct-coupled engines. With great depths and greater capacities, speed becomes a momentous factor, and direct-coupled engines are necessary. Where the depth exceeds 3,000 feet, another element enters which has given rise to much debate and experiment; that is, the great increase of starting load due to the increased length and size of ropes and the drum space required to hold it. So far the most advantageous device seems to be the Whiting hoist, a combination of double drums and tail rope.

On mines worked from near the surface, where depth is gained by the gradual exhaustion of the ore, the only prudent course is to put in a new hoist periodically, when the demand for increased winding speed and power warrants. The lack of economy in winding machines is greatly augmented if they are much over-sized for the duty. An engine installed to handle a given tonnage to a depth of 3,000 feet will have operated with more loss during the years the mine is progressing from the surface to that depth than several intermediate-sized engines would have cost. On most mines the uncertainty of extension in depth would hardly warrant such a preliminary equipment. More mines are equipped with over-sized than with under-sized engines. For shafts on going metal mines where the future is speculative, an engine will suffice whose size provides for an extension in depth of 1,000 feet beyond that reached at the time of its installation. The cost of the engine will depend more largely upon the winding speed desired than upon any other one factor. The proper speed to be arranged is obviously dependent upon the depth of the haulage, for it is useless to have an engine able to wind 3,000 feet a minute on a shaft 500 feet deep, since it could never even get under way; and besides, the relative operating loss, as said, would be enormous.

Haulage Equipment in the Shaft.—Originally, material was hoisted through shafts in buckets. Then came the cage for transporting mine cars, and in more recent years the "skip" has been developed. The aggrandized bucket or "kibble" of the Cornishman has practically disappeared, but the cage still remains in many mines. The advantages of the skip over the cage are many. Some of them are:—

It permits 25 to 40% greater load of material in proportion to the . dead weight of the vehicle.

The load can be confined within a smaller horizontal space, thus . the area of the shaft need not be so great for large tonnages.

Loading and discharging are more rapid, and the latter is automat- . ic, thus permitting more trips per hour and requiring less labor.

Skips must be loaded from bins underground, and by providing in . the bins storage capacity, shaft haulage is rendered independent of the lateral transport in the mine, and there are no delays to the engine awaiting loads. The result is that ore-winding can be concentrated into fewer hours, and indirect economies in labor and power are thus effected.

Skips save the time of the men engaged in the lateral haulage, as . they have no delay waiting for the winding engine.

Loads equivalent to those from skips are obtained in some mines by double-decked cages; but, aside from waste weight of the cage, this arrangement necessitates either stopping the engine to load the lower deck, or a double-deck loading station. Double-deck loading stations are as costly to install and more expensive to work than skip-loading station ore-bins. Cages are also constructed large enough to take as many as four trucks on one deck. This entails a shaft compartment double the size required for skips of the same capacity, and thus enormously increases shaft cost without gaining anything.

Altogether the advantages of the skip are so certain and so important that it is difficult to see the justification for the cage under but a few conditions. These conditions are those which surround mines of small output where rapidity of haulage is no object, where the cost of station-bins can thus be evaded, and the convenience of the cage for the men can still be preserved. The easy change of the skip to the cage for hauling men removes the last objection on larger mines. There occurs also the situation in which ore is broken under contract at so much per truck, and where it is desirable to inspect the contents of the truck when discharging it, but even this objection to the skip can be obviated by contracting on a cubic-foot basis.

Skips are constructed to carry loads of from two to seven tons, the general tendency being toward larger loads every year. One of the most feasible lines of improvement in winding is in the direction of larger loads and less speed, for in this way the sum total of dead weight of the vehicle and rope to the tonnage of ore hauled will be decreased, and the efficiency of the engine will be increased by a less high peak demand, because of this less proportion of dead weight and the less need of high acceleration.

LATERAL UNDERGROUND TRANSPORT.

Inasmuch as the majority of metal mines dip at considerable angles, the useful life of a roadway in a metal mine is very short because particular horizons of ore are soon exhausted. Therefore any method of transport has to be calculated upon a very quick redemption of the capital laid out. Furthermore, a roadway is limited in its daily traffic to the product of the stopes which it serves.

Men and Animals.—Some means of transport must be provided, and the basic equipment is light tracks with push-cars, in capacity from half a ton to a ton. The latter load is, however, too heavy to be pushed by one man. As but one car can be pushed at a time, hand-trucking is both slow and expensive. At average American or Australian wages, the cost works out between 25 and 35 cents a ton per mile. An improvement of growing import where hand-trucking is necessary is the overhead mono-rail instead of the track.

If the supply to any particular roadway is such as to fully employ horses or mules, the number of cars per trip can be increased up to seven or eight. In this case the expense, including wages of the men and wear, tear, and care of mules, will work out roughly at from 7 to 10 cents per ton mile. Manifestly, if the ore-supply to a particular roadway is insufficient to keep a mule busy, the economy soon runs off.

Mechanical Haulage.—Mechanical haulage is seldom applicable to metal mines, for most metal deposits dip at considerable angles, and therefore, unlike most coal-mines, the horizon of haulage must frequently change, and there are no main arteries along which haulage continues through the life of the mine. Any mechanical system entails a good deal of expense for installation, and the useful life of any particular roadway, as above said, is very short. Moreover, the crooked roadways of most metal mines present difficulties of negotiation not to be overlooked. In order to use such systems it is necessary to condense the haulage to as few roadways as possible. Where the tonnage on one level is not sufficient to warrant other than men or animals, it sometimes pays (if the dip is steep enough) to dump everything through winzes from one to two levels to a main road below where mechanical equipment can be advantageously provided. The cost of shaft-winding the extra depth is inconsiderable compared to other factors, for the extra vertical distance of haulage can be done at a cost of one or two cents per ton mile. Moreover, from such an arrangement follows the concentration of shaft-bins, and of shaft labor, and winding is accomplished without so much shifting as to horizon, all of which economies equalize the extra distance of the lift.

There are three principal methods of mechanical transport in use:—

1. Cable-ways.
2. Compressed-air locomotives.
3. Electrical haulage.

Cable-ways or endless ropes are expensive to install, and to work to the best advantage require double tracks and fairly straight roads. While they are economical in operation and work with little danger to operatives, the limitations mentioned preclude them from adoption in metal mines, except in very special circumstances such as main crosscuts or adit tunnels, where the haulage is straight and concentrated from many sources of supply.

Compressed-air locomotives are somewhat heavy and cumbersome, and therefore require well-built tracks with heavy rails, but they have very great advantages for metal mine work. They need but a single track and are of low initial cost where compressed air is already a requirement of the mine. No subsidiary line equipment is needed, and thus they are free to traverse any road in the mine and can be readily shifted from one level to another. Their mechanical efficiency is not so low in the long run as might appear from the low efficiency of pneumatic machines generally, for by storage of compressed air at the charging station a more even rate of energy consumption is possible than in the constant cable and electrical power supply which must be equal to the maximum demand, while the air-plant consumes but the average demand.

Electrical haulage has the advantage of a much more compact locomotive and the drawback of more expensive track equipment, due to the necessity of transmission wire, etc. It has the further disadvantages of uselessness outside the equipped haulage way and of the dangers of the live wire in low and often wet tunnels.

In general, compressed-air locomotives possess many attractions for metal mine work, where air is in use in any event and where any mechanical system is at all justified. Any of the mechanical systems where tonnage is sufficient in quantity to justify their employment will handle material for from 1.5 to 4 cents per ton mile.

Tracks.—Tracks for hand, mule, or rope haulage are usually built with from 12- to 16-pound rails, but when compressed-air or electrical locomotives are to be used, less than 24-pound rails are impossible. As to tracks in general, it may be said that careful laying out with even grades and gentle curves repays itself many times over in their subsequent operation. Further care in repair and lubrication of cars will often make a difference of 75% in the track resistance.

Transport in Stopes.—Owing to the even shorter life of individual stopes than levels, the actual transport of ore or waste in them is often a function of the aboriginal shovel plus gravity. As shoveling is the most costly system of transport known, any means of stoping that decreases the need for it has merit. Shrinkage-stoping eliminates it altogether. In the other methods, gravity helps in proportion to the steepness of the dip. When the underlie becomes too flat for the ore to "run," transport can sometimes be helped by pitching the ore-passes at a steeper angle than the dip (Fig. 36). In some cases of flat deposits, crosscuts into the walls, or even levels under the ore-body, are justifiable. The more numerous the ore-passes, the less the lateral shoveling, but as passes cost money for construction and for repair, there is a nice economic balance in their frequency.

Mechanical haulage in stopes has been tried and finds a field under some conditions. In dips under 25° and possessing fairly sound hanging-wall, where long-wall or flat-back cuts are employed, temporary tracks can often be laid in the stopes and the ore run in cars to the main passes. In such cases, the tracks are pushed up close to the face after each cut. Further self-acting inclines to lower cars to the levels can sometimes be installed to advantage. This arrangement also permits greater intervals between levels and less number of ore-passes. For dips between 25° and 50° where the mine is worked without stope support or with occasional pillars, a very useful contrivance is the sheet-iron trough—about eighteen inches wide and six inches deep—made in sections ten or twelve feet long and readily bolted together. In dips 35° to 50° this trough, laid on the foot-wall, gives a sufficiently smooth surface for the ore to run upon. When the dip is flat, the trough, if hung from plugs in the hanging-wall, may be swung backward and forward. The use of this "bumping-trough" saves much shoveling. For handling filling or ore in flat runs it deserves wider adoption. It is, of course, inapplicable in passes as a "bumping-trough," but can be fixed to give smooth surface. In flat mines it permits a wider interval between levels and therefore saves development work. The life of this contrivance is short when used in open stopes, owing to the dangers of bombardment from blasting.

In dips steeper than 50° much of the shoveling into passes can be saved by rill-stoping, as described on page 100. Where flat-backed stopes are used in wide ore-bodies with filling, temporary tracks laid on the filling to the ore-passes are useful, for they permit wider intervals between passes.

In that underground engineer's paradise, the Witwatersrand, where the stopes require neither timber nor filling, the long, moderately pitched openings lend themselves particularly to the swinging iron troughs, and even endless wire ropes have been found advantageous in certain cases.

Where the roof is heavy and close support is required, and where the deposits are very irregular in shape and dip, there is little hope of mechanical assistance in stope transport.

CHAPTER XIII.

MECHANICAL EQUIPMENT. (*Continued*).

DRAINAGE: CONTROLLING FACTORS; VOLUME AND HEAD OF WATER; FLEXIBILITY; RELIABILITY; POWER CONDITIONS; MECHANICAL EFFICIENCY; CAPITAL OUTLAY. SYSTEMS OF DRAINAGE,— STEAM PUMPS, COMPRESSED-AIR PUMPS, ELECTRICAL PUMPS, ROD-DRIVEN PUMPS, BAILING; COMPARATIVE VALUE OF VARIOUS SYSTEMS.

With the exception of drainage tunnels—more fully described in Chapter VIII—all drainage must be mechanical. As the bulk of mine water usually lies near the surface, saving in pumping can sometimes be effected by leaving a complete pillar of ore under some of the upper levels. In many deposits, however, the ore has too many channels to render this of much avail.

There are six factors which enter into a determination of mechanical drainage systems for metal mines:—

1. Volume and head of water.
2. Flexibility to fluctuation in volume and head.
3. Reliability.
4. Capital cost.
5. The general power conditions.
6. Mechanical efficiency.

In the drainage appliances, more than in any other feature of the equipment, must mechanical efficiency be subordinated to the other issues.

Flexibility.—Flexibility in plant is necessary because volume and head of water are fluctuating factors. In wet regions the volume of water usually increases for a certain distance with the extension of openings in depth. In dry climates it generally decreases with the downward extension of the workings after a certain depth. Moreover, as depth progresses, the water follows the openings more or less and must be pumped against an ever greater head. In most cases the volume varies with the seasons. What increase will occur, from what horizon it must be lifted, and what the fluctuations in volume are likely to be, are all unknown at the time of installation. If a pumping system were to be laid out for a new mine, which would peradventure meet every possible contingency, the capital outlay would be enormous, and the operating efficiency would be very low during the long period in which it would be working below its capacity. The question of flexibility does not arise so prominently in coal-mines, for the more or less flat deposits give a fixed factor of depth. The flow is also more steady, and the volume can be in a measure approximated from general experience.

Reliability.—The factor of reliability was at one time of more importance than in these days of high-class manufacture of many different pumping systems. Practically speaking, the only insurance from flooding in any event lies in the provision of a relief system of some sort,—duplicate pumps, or the simplest and most usual thing, bailing tanks. Only Cornish and compressed-air pumps will work with any security when drowned, and electrical pumps are easily ruined.

General Power Conditions.—The question of pumping installation is much dependent upon the power installation and other power requirements of the mine. For instance, where electrical power is purchased or generated by water-power, then electrical pumps have every advantage. Or where a large number of subsidiary motors can be economically driven from one central steam- or gas-driven electrical generation plant, they again have a strong call,—especially if the amount of water to be handled is moderate. Where the water is of limited volume and compressed-air plant a necessity for the mine, then air-driven pumps may be the most advantageous, etc.

Mechanical Efficiency.—The mechanical efficiency of drainage machinery is very largely a question of method of power application. The actual pump can be built to almost the same efficiency for any power application, and with the exception of the limited field of bailing with tanks, mechanical drainage is a matter of pumps. All pumps must be set below their load, barring a few possible feet of suction lift, and they are therefore perforce underground, and in consequence all power must be transmitted from the surface. Transmission itself means loss of power varying from 10 to 60%, depending upon the medium used. It is therefore the choice of transmission medium that largely governs the mechanical efficiency.

Systems of Drainage.—The ideal pumping system for metal mines would be one which could be built in units and could be expanded or contracted unit by unit with the fluctuation in volume; which could also be easily moved to meet the differences of lifts; and in which each independent unit could be of the highest mechanical efficiency and would require but little space for erection. Such an ideal is unobtainable among any of the appliances with which the writer is familiar.

The wide variations in the origin of power, in the form of transmission, and in the method of final application, and the many combinations of these factors, meet the demands for flexibility, efficiency, capital cost, and reliability in various degrees depending upon the environment of the mine. Power nowadays is generated primarily with steam, water, and gas. These origins admit the transmission of power to the pumps by direct steam, compressed air, electricity, rods, or hydraulic columns.

Direct Steam-pumps.—Direct steam has the disadvantage of radiated heat in the workings, of loss by the radiation, and, worse still, of the impracticability of placing and operating a highly efficient steam-engine underground. It is all but impossible to derive benefit from the vacuum, as any form of surface condenser here is impossible, and there can be no return of the hot soft water to the boilers.

Steam-pumps fall into two classes, rotary and direct-acting; the former have the great advantage of permitting the use of steam expansively and affording some field for effective use of condensation, but they are more costly, require much room, and are not fool-proof. The direct-acting pumps have all the advantage of compactness and the disadvantage of being the most inefficient of pumping machines used in mining. Taking the steam consumption of a good surface steam plant at 15 pounds per horse-power hour, the efficiency of rotary pumps with well-insulated pipes is probably not over 50%, and of direct-acting pumps from 40% down to 10%.

The advantage of all steam-pumps lies in the low capital outlay,—hence their convenient application to experimental mining and temporary pumping requirements. For final equipment they afford a great deal of flexibility, for if properly constructed they can be, with slight alteration, moved from one horizon to another without loss of relative efficiency. Thus the system can be rearranged for an increased volume of water, by decreasing the lift and increasing the number of pumps from different horizons.

Compressed-air Pumps.—Compressed-air transmission has an application similar to direct steam, but it is of still lower mechanical efficiency, because of the great loss in compression. It has the superiority of not heating the workings, and there is no difficulty as to the disposal of the exhaust, as with steam. Moreover, such pumps will work when drowned. Compressed air has a distinct place for minor pumping units, especially those removed from the shaft, for they can be run as an adjunct to the air-drill system of the mine, and by this arrangement much capital outlay may be saved. The cost of the extra power consumed by such an arrangement is less than the average cost of compressed-air power, because many of the compressor charges have to be paid anyway. When compressed air is water-generated, they have a field for permanent installations. The efficiency of even rotary air-driven pumps, based on power delivered into a good compressor, is probably not over 25%.

Electrical Pumps.—Electrical pumps have somewhat less flexibility than steam- or air-driven apparatus, in that the speed of the pumps can be varied only within small limits. They have the same great advantage in the easy reorganization of the system to altered conditions of water-flow. Electricity, when steam-generated, has the handicap of the losses of two conversions, the actual pump efficiency being about 60% in well-constructed plants; the efficiency is therefore greater than direct steam or compressed air. Where the mine is operated with water-power, purchased electric current, or where there is an installation of electrical generating plant by steam or gas for other purposes, electrically driven pumps take precedence over all others on account of their combined moderate capital outlay, great flexibility, and reasonable efficiency.

In late years, direct-coupled, electric-driven centrifugal pumps have entered the mining field, but their efficiency, despite makers' claims, is low. While they show comparatively good results on low lifts the slip increases with the lift. In heads over 200 feet their efficiency is probably not 30% of the power delivered to the electrical generator. Their chief attractions are small capital cost and the compact size which admits of easy installation.

Rod-driven Pumps.—Pumps of the Cornish type in vertical shafts, if operated to full load and if driven by modern engines, have an efficiency much higher than any other sort of installation, and records of 85 to 90% are not unusual. The highest efficiency in these pumps yet obtained has been by driving the pump with rope transmission from a high-speed triple expansion engine, and in this plant an actual consumption of only 17 pounds of steam per horse-power hour for actual water lifted has been accomplished.

To provide, however, for increase of flow and change of horizon, rod-driven pumps must be so overpowered at the earlier stage of the mine that they operate with great loss. Of all pumping systems they are the most expensive to provide. They have no place in crooked openings and only work in inclines with many disadvantages.

In general their lack of flexibility is fast putting them out of the metal miner's purview. Where the pumping depth and volume of water are approximately known, as is often the case in coal mines, this, the father of all pumps, still holds its own.

Hydraulic Pumps.—Hydraulic pumps, in which a column of water is used as the transmission fluid from a surface pump to a corresponding pump underground has had some adoption in coal mines, but little in metal mines. They have a certain amount of flexibility but low efficiency, and are not likely to have much field against electrical pumps.

Bailing.—Bailing deserves to be mentioned among drainage methods, for under certain conditions it is a most useful system, and at all times a mine should be equipped with tanks against accident to the pumps. Where the amount of water is limited,— up to, say, 50,000 gallons daily,—and where the ore output of the mine permits the use of the winding-engine for part of the time on water haulage, there is in the method an almost total saving of capital outlay. Inasmuch as the winding-engine, even when the ore haulage is finished for the day, must be under steam for handling men in emergencies, and as the labor of stokers, engine-drivers, shaft-men, etc., is therefore necessary, the cost of power consumed by bailing is not great, despite the low efficiency of winding-engines.

Comparison of Various Systems.—If it is assumed that flexibility, reliability, mechanical efficiency, and capital cost can each be divided into four figures of relative importance,—A, B, C, and D, with A representing the most desirable result,—it is possible to indicate roughly the comparative values of various pumping systems. It is not pretended that the four degrees are of equal import. In all cases the factor of general power conditions on the mine may alter the relative positions.

	DIRECT STEAM PUMPS	COMPRESSED AIR	ELECTRICITY	STEAM-DRIVEN RODS	HYDRAULIC COLUMNS	BAILING RODS
Flexibility	A	A	B	D	B	A
Reliability	B	B	B	A	D	A
Mechanical Efficiency	C	D	B	A	C	D
Capital Cost	A	B	B	D	D	—

As each mine has its special environment, it is impossible to formulate any final conclusion on a subject so involved. The attempt would lead to a discussion of a thousand supposititious cases and hypothetical remedies. Further, the description alone of pumping machines would fill volumes, and the subject will never be exhausted. The engineer confronted with pumping problems must marshal all the alternatives, count his money, and apply the tests of flexibility, reliability, efficiency, and cost, choose the system of least disadvantages, and finally deprecate the whole affair, for it is but a parasite growth on the mine.

CHAPTER XIV.

Mechanical Equipment (*Concluded*).

MACHINE DRILLING: POWER TRANSMISSION;
COMPRESSED AIR *VS*. ELECTRICITY; AIR DRILLS;
MACHINE *VS*. HAND DRILLING. WORK-SHOPS.
IMPROVEMENT IN EQUIPMENT.

For over two hundred years from the introduction of drill-holes for blasting by Caspar Weindel in Hungary, to the invention of the first practicable steam percussion drill by J. J. Crouch of Philadelphia, in 1849, all drilling was done by hand. Since Crouch's time a host of mechanical drills to be actuated by all sorts of power have come forward, and even yet the machine-drill has not reached a stage of development where it can displace hand-work under all conditions. Steam-power was never adapted to underground work, and a serviceable drill for this purpose was not found until compressed air for transmission was demonstrated by Dommeiller on the Mt. Cenis tunnel in 1861.

The ideal requirements for a drill combine:—

a. Power transmission adapted to underground conditions.
b. Lightness.
c. Simplicity of construction.
d. Strength.
e. Rapidity and strength of blow.
f. Ease of erection.
g. Reliability.
h. Mechanical efficiency.
i. Low capital cost.

No drill invented yet fills all these requirements, and all are a compromise on some point.

Power Transmission; Compressed Air *vs*. Electricity.— The only transmissions adapted to underground drill-work are compressed air and electricity, and as yet an electric-driven drill has not been produced which meets as many of the requirements of the metal miner as do compressed-air drills. The latter, up to date, have superiority in simplicity, lightness, ease of erection, reliability, and strength over electric machines. Air has another advantage in that it affords some assistance to ventilation, but it has the disadvantage of remarkably low mechanical efficiency. The actual work performed by the standard 3-3/4-inch air-drill probably does not amount to over two or three horse-power against from fifteen to eighteen horse-power delivered into the compressor, or mechanical efficiency of less than 25%. As electrical power can be delivered to the drill with much less loss than compressed air, the field for a more economical drill on this line is wide enough to create eventually the proper tool to apply it. The most satisfactory electric drill produced has been the Temple drill, which is really an air-drill driven by a small electrically-driven compressor placed near the drill itself. But even this has considerable deficiencies in mining work; the difficulties of setting up, especially for stoping work, and the more cumbersome apparatus to remove before blasting are serious drawbacks. It has deficiencies in reliability and greater complication of machinery than direct air.

Air-compression.—The method of air-compression so long accomplished only by power-driven pistons has now an alternative in some situations by the use of falling water. This latter system is a development of the last twelve years, and, due to the low initial outlay and extremely low operating costs, bids fair in those regions where water head is available not only to displace the machine compressor, but also to extend the application of compressed air to mine motors generally, and to stay in some environments the encroachment of electricity into the compressed-air field. Installations of this sort in the West Kootenay, B.C., and at the Victoria copper mine, Michigan, are giving results worthy of careful attention.

Mechanical air-compressors are steam-, water-, electrical-, and gas-driven, the alternative obviously depending on the source and cost of power. Electrical- and gas- and water-driven compressors work under the disadvantage of constant speed motors and respond little to the variation in load, a partial remedy for which lies in enlarged air-storage capacity. Inasmuch as compressed air, so far as our knowledge goes at present, must be provided for drills, it forms a convenient transmission of power to various motors underground, such as small pumps, winches, or locomotives. As stated in discussing those machines, it is not primarily a transmission of even moderate mechanical efficiency for such purposes; but as against the installation and operation of independent transmission, such as steam or electricity, the economic advantage often compensates the technical losses. Where such motors are fixed, as in pumps and winches, a considerable gain in efficiency can be obtained by reheating.

It is not proposed to enter a discussion of mechanical details of air-compression, more than to call attention to the most common delinquency in the installation of such plants. This deficiency lies in insufficient compression capacity for the needs of the mine and consequent effective operation of drills, for with under 75 pounds pressure the drills decrease remarkably in rapidity of stroke and force of the blow. The consequent decrease in actual accomplishment is far beyond the ratio that might be expected on the basis of mere difference of pressure. Another form of the same chronic ill lies in insufficient air-storage capacity to provide for maintenance of pressure against moments when all drills or motors in the mine synchronize in heavy demand for air, and thus lower the pressure at certain periods.

Air-drills.—Air-drills are from a mechanical point of view broadly of two types,—the first, in which the drill is the piston extension; and the second, a more recent development for mining work, in which the piston acts as a hammer striking the head of the drill. From an economic point of view drills may be divided into three classes. First, heavy drills, weighing from 150 to 400 pounds, which require two men for their operation; second, "baby" drills of the piston type, weighing from 110 to 150 pounds, requiring one man with occasional assistance in setting up; and third, very light drills almost wholly of the hammer type. This type is built in two forms: a heavier type for mounting on columns, weighing about 80 pounds; and a type after the order of the pneumatic riveter, weighing as low as 20 pounds and worked without mounting.

The weight and consequent mobility of a drill, aside from labor questions, have a marked effect on costs, for the lighter the drill the less difficulty and delay in erection, and consequent less loss of time and less tendency to drill holes from one radius, regardless of pointing to take best advantage of breaking planes. Moreover, smaller diameter and shorter holes consume less explosives per foot advanced or per ton broken. The best results in tonnage broken and explosive consumed, if measured by the foot of drill-hole necessary, can be accomplished from hand-drilling and the lighter the machine drill, assuming equal reliability, the nearer it approximates these advantages.

The blow, and therefore size and depth of hole and rapidity of drilling, are somewhat dependent upon the size of cylinders and length of stroke, and therefore the heavier types are better adapted to hard ground and to the deep holes of some development points. Their advantages over the other classes lie chiefly in this ability to bore exceedingly hard material and in the greater speed of advance possible in development work; but except for these two special purposes they are not as economical per foot advanced or per ton of ore broken as the lighter drills.

The second class, where men can be induced to work them one man per drill, saves in labor and gains in mobility. Many tests show great economy of the "baby" type of piston drills in average ground over the heavier machines for stoping and for most lateral development. All piston types are somewhat cumbersome and the heavier types require at least four feet of head room. The "baby" type can be operated in less space than this, but for narrow stopes they do not lend themselves with the same facility as the third class.

The third class of drills is still in process of development, but it bids fair to displace much of the occupation of the piston types of drill. Aside from being a one-man drill, by its mobility it will apparently largely reproduce the advantage of hand-drilling in ability to place short holes from the most advantageous angles and for use in narrow places. As compared with other drills it bids fair to require less time for setting up and removal and for change of bits; to destroy less steel by breakages; to dull the bits less rapidly per foot of hole; to be more economical of power; to require much less skill in operation, for judgment is less called upon in delivering speed; and to evade difficulties of fissured ground, etc. And finally the cost is only one-half, initially and for spares. Its disadvantage so far is a lack of reliability due to lightness of construction, but this is very rapidly being overcome. This type, however, is limited in depth of hole possible, for, from lack of positive reverse movement, there is a tendency for the spoil to pack around the bit, and as a result about four feet seems the limit.

The performance of a machine-drill under show conditions may be anything up to ten or twelve feet of hole per hour on rock such as compact granite; but in underground work a large proportion of the time is lost in picking down loose ore, setting up machines, removal for blasting, clearing away spoil, making adjustments, etc. The amount of lost time is often dependent upon the width of stope or shaft and the method of stoping. Situations which require long drill columns or special scaffolds greatly accentuate the loss of time. Further, the difficulties in setting up reflect indirectly on efficiency to a greater extent in that a larger proportion of holes are drilled from one radius and thus less adapted to the best breaking results than where the drill can easily be reset from various angles.

The usual duty of a heavy drill per eight-hour shift using two men is from 20 to 40 feet of hole, depending upon the rock, facilities for setting up, etc., etc.[1] The lighter drills have a less average duty, averaging from 15 to 25 feet per shift.

Machine *vs*. **Hand-Drilling.**—The advantages of hand-drilling over machine-drilling lie, first, in the total saving of power, the absence of capital cost, repairs, depreciation, etc., on

---

[1] Over the year 1907 in twenty-eight mines compiled from Alaska to Australia, an average of 23.5 feet was drilled per eight-hour shift by machines larger than three-inch cylinder.

power, compresser and drill plant; second, the time required for setting up machine-drills does not warrant frequent blasts, so that a number of holes on one radius are a necessity, and therefore machine-holes generally cannot be pointed to such advantage as hand-holes. Hand-holes can be set to any angle, and by thus frequent blasting yield greater tonnage per foot of hole. Third, a large number of comparative statistics from American, South African, and Australian mines show a saving of about 25% in explosives for the same tonnage or foot of advance by hand-holes over medium and heavy drill-holes.

The duty of a skilled white man, single-handed, in rock such as is usually met below the zone of oxidation, is from 5 to 7 feet per shift, depending on the rock and the man. Two men hand-drilling will therefore do from 1/4 to 2/3 of the same footage of holes that can be done by two men with a heavy machine-drill, and two men hand-drilling will do from 1/5 to 1/2 the footage of two men with two light drills.

The saving in labor of from 75 to 33% by machine-drilling may or may not be made up by the other costs involved in machine-work. The comparative value of machine- and hand-drilling is not subject to sweeping generalization. A large amount of data from various parts of the world, with skilled white men, shows machine-work to cost from half as much per ton or foot advanced as hand-work to 25% more than handwork, depending on the situation, type of drill, etc. In a general way hand-work can more nearly compete with heavy machines than light ones. The situations where hand-work can compete with even light machines are in very narrow stopes where drills cannot be pointed to advantage, and where the increased working space necessary for machine drills results in breaking more waste. Further, hand-drilling can often compete with machine-work in wide stopes where long columns or platforms must be used and therefore there is much delay in taking down, re-erection, etc.

Many other factors enter into a comparison, however, for machine-drilling produces a greater number of deeper holes and permits larger blasts and therefore more rapid progress. In driving levels under average conditions monthly footage is from two to three times as great with heavy machines as by hand-drilling, and by lighter machines a somewhat less proportion of greater speed. The greater speed obtained in development work, the greater tonnage obtained per man in stoping, with consequent reduction in the number of men employed, and in reduction of superintendence and general charges are indirect advantages for machine-drilling not to be overlooked.

The results obtained in South Africa by hand-drilling in shafts, and its very general adoption there, seem to indicate that better speed and more economical work can be obtained in that way in very large shafts than by machine-drilling. How far special reasons there apply to smaller shafts or labor conditions elsewhere have yet to be demonstrated. In large-dimension shafts demanding a large number of machines, the handling of long machine bars and machines generally results in a great loss of time. The large charges in deep holes break the walls very irregularly; misfires cause more delay; timbering is more difficult in the face of heavy blasting charges; and the larger amount of spoil broken at one time delays renewed drilling, and altogether the advantages seem to lie with hand-drilling in shafts of large horizontal section.

The rapid development of special drills for particular conditions has eliminated the advantage of hand-work in many situations during the past ten years, and the invention of the hammer type of drill bids fair to render hand-drilling a thing of the past. One generalization is possible, and that is, if drills are run on 40-50 pounds' pressure they are no economy over hand-drilling.

WORKSHOPS.

In addition to the ordinary blacksmithy, which is a necessity, the modern tendency has been to elaborate the shops on mines to cover machine-work, pattern-making and foundry-work, in order that delays may be minimized by quick repairs. To provide, however, for such contingencies a staff of men must be kept larger than the demand of average requirements. The result is an effort to provide jobs or to do work extravagantly or unnecessarily well. In general, it is an easy spot for fungi to start growing on the administration, and if custom repair shops are available at all, mine shops can be easily overdone.

A number of machines are now in use for sharpening drills. Machine-sharpening is much cheaper than hand-work, although the drills thus sharpened are rather less efficient owing to the difficulty of tempering them to the same nicety; however, the net results are in favor of the machines.

IMPROVEMENT IN EQUIPMENT.

Not only is every mine a progressive industry until the bottom gives out, but the technology of the industry is always progressing, so that the manager is almost daily confronted with improvements which could be made in his equipment that would result in decreasing expenses or increasing metal recovery. There is one test to the advisability of such alterations: How long will it take to recover the capital outlay from the savings effected? and over and above this recovery of capital there must be some very considerable gain. The life of mines is at least secured over the period exposed in the ore-reserves, and if the proposed alteration will show its recovery and profit in that period, then it is certainly justified. If it takes longer than this on the average speculative ore-deposit, it is a gamble on finding further ore. As a matter of practical policy it will be found that an improvement in equipment which requires more than three or four years to redeem itself out of saving, is usually a mechanical or metallurgical refinement the indulgence in which is very doubtful.

CHAPTER XV.

RATIO OF OUTPUT TO THE MINE.

DETERMINATION OF THE POSSIBLE MAXIMUM;
LIMITING FACTORS; COST OF EQUIPMENT; LIFE OF
THE MINE; MECHANICAL INEFFICIENCY OF
PATCHWORK PLANT; OVERPRODUCTION OF BASE
METAL; SECURITY OF INVESTMENT.

The output obtainable from a given mine is obviously dependent not only on the size of the deposit, but also on the equipment provided,—in which equipment means the whole working appliances, surface and underground.

A rough and ready idea of output possibilities of inclined deposits can be secured by calculating the tonnage available per foot of depth from the horizontal cross-section of the ore-bodies exposed and assuming an annual depth of exhaustion, or in horizontal deposits from an assumption of a given area of exhaustion. Few mines, at the time of initial equipment, are developed to an extent from which their possibilities in production are evident, for wise finance usually leads to the erection of some equipment and production before development has been advanced to a point that warrants a large or final installation. Moreover, even were the full possibilities of the mine known, the limitations of finance usually necessitate a less plant to start with than is finally contemplated. Therefore output and equipment are usually growing possibilities during the early life of a mine.

There is no better instance in mine engineering where pure theory must give way to practical necessities of finance than in the determination of the size of equipment and therefore output. Moreover, where finance even is no obstruction, there are other limitations of a very practical order which must dominate the question of the size of plant giving the greatest technical economy. It is, however, useful to state the theoretical considerations in determining the ultimate volume of output and therefore the size of equipments, for the theory will serve to illuminate the practical limitations. The discussion will also again demonstrate that all engineering is a series of compromises with natural and economic forces.

Output giving Least Production Cost.—As one of the most important objectives is to work the ore at the least cost per ton, it is not difficult to demonstrate that the minimum working costs can be obtained only by the most intensive production. To prove this, it need only be remembered that the working expenses of a mine are of two sorts: one is a factor of the tonnage handled, such as stoping and ore-dressing; the other is wholly or partially dependent upon time. A large number of items are of this last order. Pumping and head-office expenses are almost entirely charges independent of the tonnage handled. Superintendence and staff salaries and the like are in a large proportion dependent upon time. Many other elements of expense, such as the number of engine-drivers, etc., do not increase proportionately to increase in tonnage. These charges, or the part of them dependent upon time apart from tonnage, may be termed the "fixed charges."

There is another fixed charge more obscure yet no less certain. Ore standing in a mine is like money in a bank drawing no interest, and this item of interest may be considered a "fixed charge," for if the ore were realized earlier, this loss could be partially saved. This subject is further referred to under "Amortization."

If, therefore, the time required to exhaust the mine be prolonged by the failure to maintain the maximum output, the total cost of working it will be greater by the fixed charges over such an increased period. Conversely, by equipping on a larger scale, the mine will be exhausted more quickly, a saving in total cost can be made, and the ultimate profit can be increased by an amount corresponding to the time saved from the ravages of fixed charges. In fine, the working costs may be reduced by larger operations, and therefore the value of the mine increased.

The problem in practice usually takes the form of the relative superiority of more or of fewer units of plant, and it can be considered in more detail if the production be supposed to consist of units averaging say 100 tons per day each. The advantage of more units over less will be that the extra ones can be produced free of fixed charges, for these are an expense already involved in the lesser units. This extra production will also enjoy the interest which can be earned over the period of its earlier production. Moreover, operations on a larger scale result in various minor economies throughout the whole production, not entirely included in the type of expenditure mentioned as "fixed charges." We may call these various advantages the "saving of fixed charges" due to larger-scale operations. The saving of fixed charges amounts to very considerable sums. In general the items of working cost alone, mentioned above, which do not increase proportionately to the tonnage, aggregate from 10 to 25% of the total costs. Where much pumping is involved, the percentage will become even greater.

The question of the value of the mine as affected by the volume of output becomes very prominent in low-grade mines, where, if equipped for output on too small a scale, no profits at all could be earned, and a sufficient production is absolutely imperative for any gain. There are many mines in every country which with one-third of their present rate of production would lose money. That is, the fixed charges, if spread over small output, would be so great per ton that the profit would be extinguished by them.

In the theoretical view, therefore, it would appear clear that the greatest ultimate profit from a mine can be secured only by ore extraction under the highest pressure. As a corollary to this it follows that development must proceed with the maximum speed. Further, it follows that the present value of a mine is at least partially a factor of the volume of output contemplated.

FACTORS LIMITING THE OUTPUT.

Although the above argument can be academically defended, there are, as said at the start, practical limitations to the maximum intensity of production, arising out of many other considerations to which weight must be given. In the main, there are five principal limitations:—

1. Cost of equipment.

2. Life of the mine.
3. Mechanical inefficiency of patchwork plant.
4. Overproduction of base metal.
5. Security of investment.

Cost of Equipment.—The "saving of fixed charges" can only be obtained by larger equipment, which represents an investment. Mining works, shafts, machinery, treatment plants, and all the paraphernalia cost large sums of money. They become either worn out or practically valueless through the exhaustion of the mines. Even surface machinery when in good condition will seldom realize more than one-tenth of its expense if useless at its original site. All mines are ephemeral; therefore virtually the entire capital outlay of such works must be redeemed during the life of the mine, and the interest on it must also be recovered.

The certain life, with the exception of banket and a few other types of deposit, is that shown by the ore in sight, plus something for extension of the deposit beyond exposures. So, against the "savings" to be made, must be set the cost of obtaining them, for obviously it is of no use investing a dollar to save a total of ninety cents. The economies by increased production are, however, of such an important character that the cost of almost any number of added units (within the ability of the mine to supply them) can be redeemed from these savings in a few years. For instance, in a Californian gold mine where the working expenses are $3 and the fixed charges are at the low rate of 30 cents per ton, one unit of increased production would show a saving of over $10,000 per annum from the saving of fixed charges. In about three years this sum would repay the cost of the additional treatment equipment. If further shaft capacity were required, the period would be much extended. On a Western copper mine, where the costs are $8 and the fixed charges are 80 cents per ton, one unit of increased production would effect a saving of the fixed charges equal to the cost of the extra unit in about three years. That is, the total sum would amount to $80,000, or enough to provide almost any type of mechanical equipment for such additional tonnage.

The first result of vigorous development is to increase the ore in sight,—the visible life of the mine. When such visible life has been so lengthened that the period in which the "saving of fixed charges" will equal the amount involved in expansion of equipment, then from the standpoint of this limitation only is the added installation justified. The equipment if expanded on this practice will grow upon the heels of rapid development until the maximum production from the mine is reached, and a kind of equilibrium establishes itself.

Conversely, this argument leads to the conclusion that, regardless of other considerations, an equipment, and therefore output, should not be expanded beyond the redemption by way of "saving from fixed charges" of the visible or certain life of the mine. In those mines, such as at the Witwatersrand, where there is a fairly sound assurance of definite life, it is possible to calculate at once the size of plant which by saving of "fixed charges" will be eventually the most economical, but even here the other limitations step in to vitiate such policy of management,—chiefly the limitation through security of investment.

Life of the Mine.—If carried to its logical extreme, the above program means a most rapid exhaustion of the mine. The maximum output will depend eventually upon the rapidity with which development work may be extended. As levels and other subsidiary development openings can be prepared in inclined deposits much more quickly than the shaft can be sunk, the critical point is the shaft-sinking. As a shaft may by exertion be deepened at least 400 feet a year on a going mine, the provision of an equipment to eat up the ore-body at this rate of sinking means very early exhaustion indeed. In fact, had such a theory of production been put into practice by our forefathers, the mining profession might find difficulty in obtaining employment to-day. Such rapid exhaustion would mean a depletion of the mineral resources of the state at a pace which would be alarming.

Mechanical Inefficiency of Patchwork Plant.—Mine equipments on speculative mines (the vast majority) are often enough patchwork, for they usually grow from small beginnings; but any scheme of expansion based upon the above doctrine would need to be modified to the extent that additions could be in units large in ratio to previous installations, or their patchwork character would be still further accentuated. It would be impossible to maintain mechanical efficiency under detail expansion.

Overproduction of Base Metal.—Were this intensity of production of general application to base metal mines it would flood the markets, and, by an overproduction of metal depress prices to a point where the advantages of such large-scale operations would quickly vanish. The theoretical solution in this situation would be, if metals fell below normal prices, let the output be reduced, or let the products be stored until the price recovers. From a practical point of view either alternative is a policy difficult to face.

In the first case, reduction of output means an increase of working expenses by the spread of fixed charges over less tonnage, and this in the face of reduced metal prices. It may be contended, however, that a falling metal market is usually the accompaniment of a drop in all commodities, wherefore working costs can be reduced somewhat in such times of depression, thereby partially compensating the other elements making for increased costs. Falls in commodities are also the accompaniment of hard times. Consideration of one's workpeople and the wholesale slaughter of dividends to the then needy stockholders, resulting from a policy of reduced production, are usually sufficient deterrents to diminished output.

The second alternative, that of storing metal, means equally a loss of dividends by the investment of a large sum in unrealized products, and the interest on this sum. The detriment to the market of large amounts of unsold metal renders such a course not without further disadvantages.

Security of Investment.—Another point of view antagonistic to such wholesale intensity of production, and one worthy of careful consideration, is that of the investor in mines. The root-value of mining stocks is, or should be, the profit in sight. If the policy of greatest economy in production costs be followed as outlined above, the economic limit of ore-reserves gives an apparently very short life, for the ore in sight will never represent a life beyond the time required to justify more plant. Thus the "economic limit of ore in reserve" will be a store equivalencing a period during which additional equipment can be redeemed from the "saving of fixed charges," or three or four years, usually.

The investor has the right to say that he wants the guarantee of longer life to his investment,—he will in effect pay insurance for it by a loss of some ultimate profit. That this view, contradictory to the economics of the case, is not simply academic, can be observed by any one who studies what mines are in best repute on any stock exchange. All engineers must wish to have the industry under them in high repute. The writer knows of several mines paying 20% on their stocks which yet stand lower in price on account of short ore-reserves than mines paying less annual returns. The speculator, who is an element not to be wholly disregarded, wishes a rise in his mining stock, and if development proceeds at a pace in advance of production, he will gain a legitimate rise through the increase in ore-reserves.

The investor's and speculator's idea of the desirability of a proved long life readily supports the technical policy of high-pressure development work, but not of expansion of production, for they desire an increasing ore-reserve. Even the metal operator who is afraid of overproduction does not object to increased ore-reserves. On the point of maximum intensity of development work in a mine all views coincide. The mining engineer, if he takes a Machiavellian view, must agree with the investor and the metal dealer, for the engineer is a "fixed charge" the continuance of which is important to his daily needs.

The net result of all these limitations is therefore an invariable compromise upon some output below the possible maximum. The initial output to be contemplated is obviously one upon which the working costs will be low enough to show a margin of profit. The medium between these two extremes is determinable by a consideration of the limitations set out,—and the cash available. When the volume of output is once determined, it must be considered as a factor in valuation, as discussed under "Amortization."

CHAPTER XVI.

ADMINISTRATION.

LABOR EFFICIENCY; SKILL; INTELLIGENCE;
APPLICATION COORDINATION; CONTRACT WORK;
LABOR UNIONS; REAL BASIS OF WAGES.

The realization from a mine of the profits estimated from the other factors in the case is in the end dependent upon the management. Good mine management is based upon three elementals: first, sound engineering; second, proper co-ordination and efficiency of every human unit; third, economy in the purchase and consumption of supplies.

The previous chapters have been devoted to a more or less extended exposition of economic engineering. While the second and third requirements are equally important, they range in many ways out of the engineering and into the human field. For this latter reason no complete manual will ever be published upon "How to become a Good Mine Manager."

It is purposed, however, to analyze some features of these second and third fundamentals, especially in their interdependent phases, and next to consider the subject of mine statistics, for the latter are truly the microscopes through which the competence of the administration must be examined.

The human units in mine organization can be divided into officers and men. The choice of mine officers is the assembling of specialized brains. Their control, stimulation, and inspiration is the main work of the administrative head. Success in the selection and control of staff is the index of executive ability. There are no mathematical, mechanical, or chemical formulas for dealing with the human mind or human energies.

Labor.—The whole question of handling labor can be reduced to the one term "efficiency." Not only does the actual labor outlay represent from 60 to 70% of the total underground expenses, but the capacity or incapacity of its units is responsible for wider fluctuations in production costs than the bare predominance in expenditure might indicate. The remaining expense is for supplies, such as dynamite, timber, steel, power, etc., and the economical application of these materials by the workman has the widest bearing upon their consumption.

Efficiency of the mass is the resultant of that of each individual under a direction which co-ordinates effectively all units. The lack of effectiveness in one individual diminishes the returns not simply from that man alone; it lowers the results from numbers of men associated with the weak member through the delaying and clogging of their work, and of the machines operated by them. Co-ordination of work is a necessary factor of final efficiency. This is a matter of organization and administration. The most zealous stoping-gang in the world if associated with half the proper number of truckers must fail to get the desired result.

Efficiency in the single man is the product of three factors,—skill, intelligence, and application. A great proportion of underground work in a mine is of a type which can be performed after a fashion by absolutely unskilled and even unintelligent men, as witness the breaking-in of savages of low average mentality, like the South African Kaffirs. Although most duties can be performed by this crudest order of labor, skill and intelligence can be applied to it with such economic results as to compensate for the difference in wage. The reason for this is that the last fifty years have seen a substitution of labor-saving machines for muscle. Such machines displace hundreds of raw laborers. Not only do they initially cost large sums, but they require large expenditure for power and up-keep. These fixed charges against the machine demand that it shall be worked at its maximum. For interest, power, and up-keep go on in any event, and the saving on crude labor displaced is not so great but that it quickly disappears if the machine is run under its capacity. To get its greatest efficiency, a high degree of skill and intelligence is required. Nor are skill and intelligence alone applicable to labor-saving devices themselves, because drilling and blasting rock and executing other works underground are matters in which experience and judgment in the individual workman count to the highest degree.

How far skill affects production costs has had a thorough demonstration in West Australia. For a time after the opening of those mines only a small proportion of experienced men were obtainable. During this period the rock broken per man employed underground did not exceed the rate of 300 tons a year. In the large mines it has now, after some eight years, attained 600 to 700 tons.

How far intelligence is a factor indispensable to skill can be well illustrated by a comparison of the results obtained from working labor of a low mental order, such as Asiatics and negroes, with those achieved by American or Australian miners. In a general way, it may be stated with confidence that the white miners above mentioned can, under the same physical conditions, and with from five to ten times the wage, produce the same economic result,—that is, an equal or lower cost per unit of production. Much observation and experience in working Asiatics and negroes as well as Americans and Australians in mines, leads the writer to the conclusion that, averaging actual results, one white man equals from two to three of the colored races, even in the simplest forms of mine work such as shoveling or tramming. In the most highly skilled branches, such as mechanics, the average ratio is as one to seven, or in extreme cases even eleven. The question is not entirely a comparison of bare efficiency individually; it is one of the sum total of results. In mining work the lower races require a greatly increased amount of direction, and this excess of supervisors consists of men not in themselves directly productive. There is always, too, a waste of supplies, more accidents, and more ground to be kept open for accommodating increased staff, and the maintenance of these openings must be paid for. There is an added expense for handling larger numbers in and out of the mine, and the lower intelligence reacts in many ways in lack of co-ordination and inability to take initiative. Taking all divisions of labor together, the ratio of efficiency as measured in amount of output works out from four to five colored men as the equivalent of one white man of the class stated. The ratio of costs, for reasons already mentioned, and in other than quantity relation, figures still more in favor of the higher intelligence.

The following comparisons, which like all mine statistics must necessarily be accepted with reservation because of some dissimilarity of economic surroundings, are yet on sufficiently common ground to demonstrate the main issue,—that is, the bearing of inherent intelligence in the workmen and their consequent skill. Four groups of gold mines have been taken, from India, West Australia, South Africa, and Western America. All of those chosen are of the same stoping width, 4 to 5 feet. All are working in depth and with every labor-saving device available. All dip at about the same angle and are therefore in much the same position as to handling rock. The other conditions are against the white-manned mines and in favor of the colored. That is, the Indian mines have water-generated electric power and South Africa has cheaper fuel than either the American or Australian examples. In both the white-manned groups, the stopes are supported, while in the others no support is required.

GROUP OF MINES	TONS OF MATERIAL EXCAVATED OVER PERIOD COMPILED[1]	AVERAGE NUMBER OF MEN EMPLOYED		TONS PER MAN PER ANNUM	COST PER TON OF MATERIAL BROKEN
		Colored	White		
Four Kolar mines[2]	963,950	13,611	302	69.3	$3.85
Six Australian mines[3]	1,027,718	—	1,534	669.9	2.47
Three Witwatersrand mines[4]	2,962,640	13,560	1,595	195.5	2.68
Five American mines[1]	1,089,500	—	1,524	713.3	1.92

[1] Includes rock broken in development work. In the case of the specified African mines, the white labor is employed almost wholly in positions of actual or semi-superintendence, such as one white man in charge of two or three drills. In the Indian case, in addition to the white men who are wholly in superintendence, there were of the natives enumerated some 1000 in positions of semi-superintendence, as contractors or headmen, working-gangers, etc.
[2] Indian wages average about 20 cents per day.
[3] White men's wages average about $3 per day.
[4] About two-fifths of the colored workers were negroes, and three-fifths Chinamen. Negroes average about 60 cents, and Chinamen about 45 cents per day, including keep.

One issue arises out of these facts, and that is that no engineer or investor in valuing mines is justified in anticipating lower costs in regions where cheap labor exists.

In supplement to sheer skill and intelligence, efficiency can be gained only by the application of the man himself. A few months ago a mine in California changed managers. The new head reduced the number employed one-third without impairing the amount of work accomplished. This was not the result of higher skill or intelligence in the men, but in the manager. Better application and co-ordination were secured from the working force. Inspiration to increase of exertion is created less by "driving" than by recognition of individual effort, in larger pay, and by extending justifiable hope of promotion. A great factor in the proficiency of the mine manager is his ability to create an *esprit-de-corps* through the whole staff, down to the last tool boy. Friendly interest in the welfare of the men and stimulation by competitions between various works and groups all contribute to this end.

Contract Work.—The advantage both to employer and employed of piece work over wage needs no argument. In a general way, contract work honorably carried out puts a premium upon individual effort, and thus makes for efficiency. There are some portions of mine work which cannot be contracted, but the development, stoping, and trucking can be largely managed in this way, and these items cover 65 to 75% of the total labor expenditure underground.

In development there are two ways of basing contracts,—the first on the footage of holes drilled, and the second on the footage of heading advanced. In contract-stoping there are four methods depending on the feet of hole drilled, on tonnage, on cubic space, and on square area broken.

All these systems have their rightful application, conditioned upon the class of labor and character of the deposit.

In the "hole" system, the holes are "pointed" by some mine official and are blasted by a special crew. The miner therefore has little interest in the result of the breaking. If he is a skilled

[1] Wages about $3.50. Tunnel entry in two mines.

white man, the hours which he has wherein to contemplate the face usually enable him to place holes to better advantage than the occasional visiting foreman. With colored labor, the lack of intelligence in placing holes and blasting usually justifies contracts per "foot drilled." Then the holes are pointed and blasted by superintending men.

On development work with the foot-hole system, unless two working faces can be provided for each contracting party, they are likely to lose time through having finished their round of holes before the end of the shift. As blasting must be done outside the contractor's shifts, it means that one shift per day must be set aside for the purpose. Therefore not nearly such progress can be made as where working the face with three shifts. For these reasons, the "hole" system is not so advantageous in development as the "foot of advance" basis.

In stoping, the "hole" system has not only a wider, but a sounder application. In large ore-bodies where there are waste inclusions, it has one superiority over any system of excavation measurement, namely, that the miner has no interest in breaking waste into the ore.

The plan of contracting stopes by the ton has the disadvantage that either the ore produced by each contractor must be weighed separately, or truckers must be trusted to count correctly, and to see that the cars are full. Moreover, trucks must be inspected for waste,—a thing hard to do underground. So great are these detailed difficulties that many mines are sending cars to the surface in cages when they should be equipped for bin-loading and self-dumping skips.

The method of contracting by the cubic foot of excavation saves all necessity for determining the weight of the output of each contractor. Moreover, he has no object in mixing waste with the ore, barring the breaking of the walls. This system therefore requires the least superintendence, permits the modern type of hoisting, and therefore leaves little justification for the survival of the tonnage basis.

Where veins are narrow, stoping under contract by the square foot or fathom measured parallel to the walls has an advantage. The miner has no object then in breaking wall-rock, and the thoroughness of the ore-extraction is easily determined by inspection.

Bonus Systems.—By giving cash bonuses for special accomplishment, much the same results can be obtained in some departments as by contracting. A bonus per foot of heading gained above a minimum, or an excess of trucks trammed beyond a minimum, or prizes for the largest amount done during the week or month in special works or in different shifts,—all these have a useful application in creating efficiency. A high level of results once established is easily maintained.

Labor Unions.—There is another phase of the labor question which must be considered and that is the general relations of employer and employed. In these days of largely corporate proprietorship, the owners of mines are guided in their relations with labor by engineers occupying executive positions. On them falls the responsibility in such matters, and the engineer becomes thus a buffer between labor and capital. As corporations have grown, so likewise have the labor unions. In general, they are normal and proper antidotes for unlimited capitalistic organization.

Labor unions usually pass through two phases. First, the inertia of the unorganized labor is too often stirred only by demagogic means. After organization through these and other agencies, the lack of balance in the leaders often makes for injustice in demands, and for violence to obtain them and disregard of agreements entered upon. As time goes on, men become educated in regard to the rights of their employers, and to the reflection of these rights in ultimate benefit to labor itself. Then the men, as well as the intelligent employer, endeavor to safeguard both interests. When this stage arrives, violence disappears in favor of negotiation on economic principles, and the unions achieve their greatest real gains. Given a union with leaders who can control the members, and who are disposed to approach differences in a business spirit, there are few sounder positions for the employer, for agreements honorably carried out dismiss the constant harassments of possible strikes. Such unions exist in dozens of trades in this country, and they are entitled to greater recognition. The time when the employer could ride roughshod over his labor is disappearing with the doctrine of "*laissez faire*," on which it was founded. The sooner the fact is recognized, the better for the employer. The sooner some miners' unions develop from the first into the second stage, the more speedily will their organizations secure general respect and influence.[1]

The crying need of labor unions, and of some employers as well, is education on a fundamental of economics too long disregarded by all classes and especially by the academic economist. When the latter abandon the theory that wages are the result of supply and demand, and recognize that in these days of international flow of labor, commodities and capital, the real controlling factor in wages is efficiency, then such an educational campaign may become possible. Then will the employer and employee find a common ground on which each can benefit. There lives no engineer who has not seen insensate dispute as to wages where the real difficulty was inefficiency. No administrator begrudges a division with his men of the increased profit arising from increased efficiency. But every administrator begrudges the wage level demanded by labor unions whose policy is decreased efficiency in the false belief that they are providing for more labor.

[1] Some years of experience with compulsory arbitration in Australia and New Zealand are convincing that although the law there has many defects, still it is a step in the right direction, and the result has been of almost unmixed good to both sides. One of its minor, yet really great, benefits has been a considerable extinction of the parasite who lives by creating violence.

CHAPTER XVII.

ADMINISTRATION (*Continued*).

ACCOUNTS AND TECHNICAL DATA AND REPORTS;
WORKING COSTS; DIVISION OF EXPENDITURE;
INHERENT LIMITATIONS IN ACCURACY OF
WORKING COSTS; WORKING COST SHEETS.
GENERAL TECHNICAL DATA; LABOR, SUPPLIES,
POWER, SURVEYS, SAMPLING, AND ASSAYING.

First and foremost, mine accounts are for guidance in the distribution of expenditure and in the collection of revenue; secondly, they are to determine the financial progress of the enterprise, its profit or loss; and thirdly, they are to furnish statistical data to assist the management in its interminable battle to reduce expenses and increase revenue, and to enable the owner to determine the efficiency of his administrators. Bookkeeping *per se* is no part of this discussion. The fundamental purpose of that art is to cover the first two objects, and, as such, does not differ from its application to other commercial concerns.

In addition to these accounting matters there is a further type of administrative report of equal importance—that is the periodic statements as to the physical condition of the property, the results of exploration in the mine, and the condition of the equipment.

ACCOUNTS.

The special features of mine accounting reports which are a development to meet the needs of this particular business are the determination of working costs, and the final presentation of these data in a form available for comparative purposes.

The subject may be discussed under:—
1. Classes of mine expenditure.
2. Working costs.

3. The dissection of expenditures departmentally.
4. Inherent limitations in the accuracy of working costs.
5. Working cost sheets.

In a wide view, mine expenditures fall into three classes, which maybe termed the "fixed charges," "proportional charges," and "suspense charges" or "capital expenditure." "Fixed charges" are those which, like pumping and superintendence, depend upon time rather than tonnage and material handled. They are expenditures that would not decrease relatively to output. "Proportional charges" are those which, like ore-breaking, stoping, supporting stopes, and tramming, are a direct coefficient of the ore extracted. "Suspense charges" are those which are an indirect factor of the cost of the ore produced, such as equipment and development. These expenditures are preliminary to output, and they thus represent a storage of expense to be charged off when the ore is won. This outlay is often called "capital expenditure." Such a term, though in common use, is not strictly correct, for the capital value vanishes when the ore is extracted, but in conformity with current usage the term "capital expenditure" will be adopted.

Except for the purpose of special inquiry, such as outlined under the chapter on "Ratio of Output," "fixed charges" are not customarily a special division in accounts. In a general way, such expenditures, combined with the "proportional charges," are called "revenue expenditure," as distinguished from the capital, or "suspense," expenditures. In other words, "revenue" expenditures are those involved in the daily turnover of the business and resulting in immediate returns. The inherent difference in character of revenue and capital expenditures is responsible for most of the difficulties in the determination of working costs, and most of the discussion on the subject.

Working Costs.—"Working costs" are a division of expenditure for some unit,—the foot of opening, ton of ore, a pound of metal, cubic yard or fathom of material excavated, or some other measure. The costs per unit are usually deduced for each month and each year. They are generally determined for each of the different departments of the mine or special works separately. Further, the various sorts of expenditure in these departments are likewise segregated.

In metal mining the ton is the universal unit of distribution for administrative purpose, although the pound of metal is often used to indicate final financial results. The object of determination of "working costs" is fundamentally for comparative purposes. Together with other technical data, they are the nerves of the administration, for by comparison of detailed and aggregate results with other mines and internally in the same mine, over various periods and between different works, a most valuable check on efficiency is possible. Further, there is one collateral value in all statistical data not to be overlooked, which is that the knowledge of its existence induces in the subordinate staff both solicitude and emulation.

The fact must not be lost sight of, however, that the wide variations in physical and economic environment are so likely to vitiate conclusions from comparisons of statistics from two mines or from two detailed works on the same mine, or even from two different months on the same work, that the greatest care and discrimination are demanded in their application. Moreover, the inherent difficulties in segregating and dividing the accounts which underlie such data, render it most desirable to offer some warning regarding the limits to which segregation and division may be carried to advantage.

As working costs are primarily for comparisons, in order that they may have value for this purpose they must include only such items of expenditure as will regularly recur. If this limitation were more generally recognized, a good deal of dispute and polemics on the subject might be saved. For this reason it is quite impossible that all the expenditure on the mine should be charged into working costs, particularly some items that arise through "capital expenditure."

The Dissection of Expenditures Departmentally.—The final division in the dissection of the mine expenditure is in the main:—

Revenue
- (1 General Expenses
- (2 Ore Extraction — Ore-breaking. Supporting Stopes. Trucking Ore. Hoisting.
- (3 Pumping

Capital or Suspense.
- (4 Development — Shaft-sinking. Station-cutting. Crosscutting. Driving. Rising. Winzes. Diamond Drilling.
- (5 Construction and Equipment — Various Works

Various expenditures for labor, supplies, power, repairs, etc., worked out per ton or foot advanced over each department

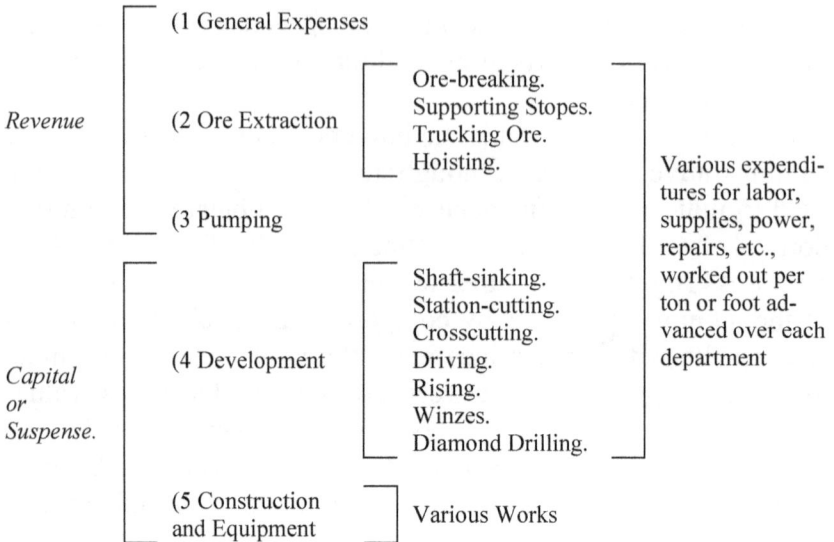

The detailed dissection of expenditures in these various departments with view to determine amount of various sorts of expenditure over the department, or over some special work in that department, is full of unsolvable complications. The allocation of the direct expenditure of labor and supplies applied to the above divisions or special departments in them, is easily accomplished, but beyond this point two sorts of difficulties immediately arise and offer infinite field for opinion and method. The first of these difficulties arises from supplementary departments on the mine, such as "power," "repairs and maintenance," "sampling and assaying." These departments must be "spread" over the divisions outlined above, for such charges are in part or whole a portion of the expense of these divisions. Further, all of these "spread" departments are applied to surface as well as to underground works, and must be divided not only over the above departments but also over the surface departments,— not under discussion here. The common method is to distribute "power" on a basis of an approximation of the amount used in each department; to distribute "repairs and maintenance," either on a basis of shop returns, or a distribution over all departments on the basis of the labor employed in those departments, on the theory that such repairs arise in this proportion; to distribute sampling and assaying over the actual points to which they relate at the average cost per sample or assay.

"General expenses," that is, superintendence, etc., are often not included in the final departments as above, but are sometimes "spread" in an attempt to charge a proportion of superintendence to each particular work. As, however, such "spreading" must take place on the basis of the relative expenditure in each department, the result is of little value, for such a basis does not truly represent the proportion of general superintendence, etc., devoted to each department. If they are distributed over all departments, capital as well as revenue, on the basis of total expenditure, they inflate the "capital expenditure" departments against a day of reckoning when these charges come to be distributed over working costs. Although it may be contended that the capital departments also require supervision, such a practice is a favorite device for showing apparently low working costs in the revenue departments. The most courageous way is not to distribute general expenses at all, but to charge them separately and directly to revenue accounts and thus wholly into working costs.

The second problem is to reduce the "suspense" or capital charges to a final cost per ton, and this is no simple matter. Development expenditures bear a relation to the tonnage developed and not to that extracted in any particular period. If it is desired to preserve any value for comparative purposes in the mining costs, such outlay must be charged out on the basis of the tonnage developed, and such portion of the ore as is extracted must be written off at this rate; otherwise one month may see double the amount of development in progress which another records, and the underground costs would be swelled or diminished thereby in a way to ruin their comparative value from month to month. The ore developed cannot be satisfactorily determined at short intervals, but it can be known at least annually, and a price may be deduced as to its cost per ton. In many mines a figure is arrived at by estimating ore-reserves at the end of the year, and this figure is used during the succeeding year as a "redemption of development" and as such charged to working costs, and thus into revenue account in proportion to the tonnage extracted. This matter is further elaborated in some mines, in that winzes and rises are written off at one rate, levels and crosscuts at another, and shafts at one still lower, on the theory that they lost their usefulness in this progression as the ore is extracted. This course, however, is a refinement hardly warranted.

Plant and equipment constitute another "suspense" account even harder to charge up logically to tonnage costs, for it is in many items dependent upon the life of the mine, which is an unknown factor. Most managers debit repairs and maintenance directly to the revenue account and leave the reduction of the construction outlay to an annual depreciation on the final balance sheet, on the theory that the plant is maintained out of costs to its original value. This subject will be discussed further on.

Inherent Limitations in Accuracy of Working Costs.— There are three types of such limitations which arise in the determination of costs and render too detailed dissection of such costs hopeless of accuracy and of little value for comparative purposes. They are, first, the difficulty of determining all of even direct expenditure on any particular crosscut, stope, haulage, etc.; second, the leveling effect of distributing the "spread" expenditures, such as power, repairs, etc.; and third, the difficulties arising out of the borderland of various departments.

Of the first of these limitations the instance may be cited that foremen and timekeepers can indicate very closely the destination of labor expense, and also that of some of the large items of supply, such as timber and explosives, but the distribution of minor supplies, such as candles, drills, picks, and shovels, is impossible of accurate knowledge without an expense wholly unwarranted by the information gained. To determine at a particular crosscut the exact amount of steel, and of tools consumed, and the cost of sharpening them, would entail their separate and special delivery to the same place of attack and a final weighing-up to learn the consumption.

Of the second sort of limitations, the effect of "spread" expenditure, the instance may be given that the repairs and maintenance are done by many men at work on timbers, tracks, machinery, etc. It is hopeless to try and tell how much of their work should be charged specifically to detailed points. In the distribution of power may be taken the instance of air-drills. Although the work upon which the drill is employed can be known, the power required for compression usually comes from a common power-plant, so that the portion of power debited to the air compressor is an approximation. The assumption of an equal consumption of air by all drills is a further approximation. In practice, therefore, many expenses are distributed on the theory that they arise in proportion to the labor employed, or the machines used in the various departments. The net result is to level down expensive points and level up inexpensive ones.

The third sort of limitation of accounting difficulty referred to, arises in determining into which department are actually to be allocated the charges which lie in the borderland between various primary classes of expenditure. For instance, in ore won from development,—in some months three times as much development may be in ore as in other months. If the total expense of development work which yields ore be charged to stoping account, and if cost be worked out on the total tonnage of ore hoisted, then the stoping cost deduced will be erratic, and the true figures will be obscured. On the other hand, if all development is charged to 'capital account' and the stoping cost worked out on all ore hoisted, it will include a fluctuating amount of ore not actually paid for by the revenue departments or charged into costs. This fluctuation either way vitiates the whole comparative value of the stoping costs. In the following system a compromise is reached by crediting "development" with an amount representing the ore won from development at the average cost of stoping, and by charging this amount into "stoping." A number of such questions arise where the proper division is simply a matter of opinion.

The result of all these limitations is that a point in detail is quickly reached where no further dissection of expenditure is justified, since it becomes merely an approximation. The writer's own impression is that without an unwarrantable number of accountants, no manager can tell with any accuracy the cost of any particular stope, or of any particular development heading. Therefore, aside from some large items, such detailed statistics, if given, are to be taken with great reserve.

Working Cost Sheets.—There are an infinite number of forms of working cost sheets, practically every manager having a system of his own. To be of greatest value, such sheets should show on their face the method by which the "spread" departments are handled, and how revenue and suspense departments are segregated. When too much detail is presented, it is but a waste of accounting and consequent expense. Where to draw the line in this regard is, however, a matter of great difficulty. No cost sheet is entirely satisfactory. The appended sheet is in use at a number of mines. It is no more perfect than many others. It will be noticed that the effect of this system is to throw the general expenses into the revenue expenditures, and as little as possible into the "suspense" account.

GENERAL TECHNICAL DATA.

For the purposes of efficient management, the information gathered under this head is of equal, if not superior, importance to that under "working costs." Such data fall generally under the following heads:—

Labor.—Returns of the shifts worked in the various departments for each day and for the month; worked out on a monthly basis of footage progress, tonnage produced or tons handled per man; also where possible the footage of holes drilled, worked out per man and per machine.

Supplies.—Daily returns of supplies used; the principal items worked out monthly in quantity per foot of progress, or per ton of ore produced.

Power.—Fuel, lubricant, etc., consumed in steam production, worked out into units of steam produced, and this production allocated to the various engines. Where electrical power is used, the consumption of the various motors is set out.

Surveys.—The need of accurate plans requires no discussion. Aside from these, the survey-office furnishes the returns of development footage, measurements under contracts, and the like.

Sampling and Assaying.—Mine sampling and assaying fall under two heads,—the determination of the value of standing ore, and of products from the mine. The sampling and assaying on a going mine call for the same care and method as in cases of valuation of the mine for purchase,—the details of which have been presented under "Mine Valuation,"—for through it, guidance must not only be had to the value of the mine and for reports to owners, but the detailed development and ore extraction depend on an absolute knowledge of where the values lie.

CHAPTER XVIII.

ADMINISTRATION (*Concluded*).

ADMINISTRATIVE REPORTS.

In addition to financial returns showing the monthly receipts, expenditures, and working costs, there must be in proper administration periodic reports from the officers of the mine to the owners or directors as to the physical progress of the enterprise. Such reports must embrace details of ore extraction, metal contents, treatment recoveries, construction of equipment, and the results of underground development. The value of mines is so much affected by the monthly or even daily result of exploration that reports of such work are needed very frequently,—weekly or even daily if critical work is in progress. These reports must show the width, length, and value of the ore disclosed.

The tangible result of development work is the tonnage and grade of ore opened up. How often this stock-taking should take place is much dependent upon the character of the ore. The result of exploration in irregular ore-bodies often does not, over short periods, show anything tangible in definite measurable tonnage, but at least annually the ore reserve can be estimated.

In mines owned by companies, the question arises almost daily as to how much of and how often the above information should be placed before stockholders (and therefore the public) by the directors. In a general way, any company whose shares are offered on the stock exchange is indirectly inviting the public to become partners in the business, and these partners are entitled to all the information which affects the value of their property and are entitled to it promptly. Moreover, mining is a business where competition is so obscure and so much a matter of indifference, that suppression of important facts in documents for public circulation has no justification. On the other hand, both the technical progress of the industry and its position in public esteem demand the fullest disclosure and greatest care in preparation of reports. Most stockholders' ignorance of mining technology and of details of their particular mine demands a great deal of care and discretion in the preparation of these public reports that they may not be misled. Development results may mean little or much, depending upon the location of the work done in relation to the ore-bodies, etc., and this should be clearly set forth.

The best opportunity of clear, well-balanced statements lies in the preparation of the annual report and accounts. Such reports are of three parts:—

1. The "profit and loss" account, or the "revenue account."
2. The balance sheet; that is, the assets and liabilities statement.
3. The reports of the directors, manager, and consulting engineer.

The first two items are largely matters of bookkeeping. They or the report should show the working costs per ton for the year. What must be here included in costs is easier of determination than in the detailed monthly cost sheets of the administration; for at the annual review, it is not difficult to assess the amount chargeable to development. Equipment expenditure, however, presents an annual difficulty, for, as said, the distribution of this item is a factor of the life of the mine, and that is unknown. If such a plant has been paid for out of the earnings, there is no object in carrying it on the company's books as an asset, and most well-conducted companies write it off at once. On the other hand, where the plant is paid for out of capital provided for the purpose, even to write off depreciation means that a corresponding sum of cash must be held in the company's treasury in order to balance the accounts,—in other words, depreciation in such an instance becomes a return of capital. The question then is one of policy in the company's finance, and in neither case is it a matter which can be brought into working costs and leave them any value for comparative purposes. Indeed, the true cost of working the ore from any mine can only be told when the mine is exhausted; then the dividends can be subtracted from the capital sunk and metal sold, and the difference divided over the total tonnage produced.

The third section of the report affords wide scope for the best efforts of the administration. This portion of the report falls into three divisions: (*a*) the construction and equipment work of the year, (*b*) the ore extraction and treatment, and (*c*) the results of development work.

The first requires a statement of the plant constructed, its object and accomplishment; the second a disclosure of tonnage produced, values, metallurgical and mechanical efficiency. The third is of the utmost importance to the stockholder, and is the one most often disregarded and obscured. Upon this hinges the value of the property. There is no reason why, with plans and simplicity of terms, such reports cannot be presented in a manner from which the novice can judge of the intrinsic position of the property. A statement of the tonnage of ore-reserves and their value, or of the number of years' supply of the current output, together with details of ore disclosed in development work, and the working costs, give the ground data upon which any stockholder who takes interest in his investment may judge for himself. Failure to provide such data will some day be understood by the investing public as a *prima facie* index of either incapacity or villainy. By the insistence of the many engineers in administration of mines upon the publication of such data, and by the insistence of other engineers upon such data for their clients before investment, and by the exposure of the delinquents in the press, a more practicable "protection of investors" can be reached than by years of academic discussion.

CHAPTER XIX.

THE AMOUNT OF RISK IN MINING INVESTMENTS.

RISK IN VALUATION OF MINES; IN MINES AS
COMPARED WITH OTHER COMMERCIAL
ENTERPRISES.

From the constant reiteration of the risks and difficulties involved in every step of mining enterprise from the valuation of
the mine to its administration as a going concern, the impression
may be gained that the whole business is one great gamble; in
other words, that the point whereat certainties stop and conjec-
ture steps in is so vital as to render the whole highly speculative.

Far from denying that mining is, in comparison with better-
class government bonds, a speculative type of investment, it is
desirable to avow and emphasize the fact. But it is none the less
well to inquire what degree of hazard enters in and how it com-
pares with that in other forms of industrial enterprise.

Mining business, from an investment view, is of two sorts,—
prospecting ventures and developed mines; that is, mines where
little or no ore is exposed, and mines where a definite quantity of
ore is measurable or can be reasonably anticipated. The great
hazards and likewise the Aladdin caves of mining are mainly
confined to the first class. Although all mines must pass through
the prospecting stage, the great industry of metal production is
based on developed mines, and it is these which should come
into the purview of the non-professional investor. The first class
should be reserved invariably for speculators, and a speculator
may be defined as one who hazards all to gain much. It is with
mining as an investment, however, that this discussion is concerned.

Risk in Valuation of Mines.—Assuming a competent collection of data and efficient management of the property, the risks
in valuing are from step to step:—

1. The risk of continuity in metal contents beyond sample faces.
2. The risk of continuity in volume through the blocks estimated.
3. The risk of successful metallurgical treatment.
4. The risk of metal prices, in all but gold.
5. The risk of properly estimating costs.
6. The risk of extension of the ore beyond exposures.
7. The risk of management.

As to the continuity of values and volumes through the estimated area, the experience of hundreds of engineers in hundreds of mines has shown that when the estimates are based on properly secured data for "proved ore," here at least there is absolutely no hazard. Metallurgical treatment, if determined by past experience on the ore itself, carries no chance; and where determined by experiment, the risk is eliminated if the work be sufficiently exhaustive. The risk of metal price is simply a question of how conservative a figure is used in estimating. It can be eliminated if a price low enough be taken. Risk of extension in depth or beyond exposures cannot be avoided. It can be reduced in proportion to the distance assumed. Obviously, if no extension is counted, there is nothing chanced. The risk of proper appreciation of costs is negligible where experience in the district exists. Otherwise, it can be eliminated if a sufficiently large allowance is taken. The risk of failure to secure good management can be eliminated if proved men are chosen.

There is, therefore, a basic value to every mine. The "proved" ore taken on known metallurgical grounds, under known conditions of costs on minimum prices of metals, has a value as certain as that of money in one's own vault. This is the value previously referred to as the "A" value. If the price (and interest on it pending recovery) falls within this amount, there is no question that the mine is worth the price. What the risk is in mining is simply what amount the price of the investment demands shall be won from extension of the deposit beyond known exposures, or what higher price of metal must be realized than that calculated in the "A" value. The demands on this $X,\ Y$ portion of the mine can be converted into tons of ore, life of production, or higher prices, and these can be weighed with the geological weights and the industrial outlook.

Mines compared to Other Commercial Enterprises.—The profits from a mining venture over and above the bed-rock value A, that is, the return to be derived from more extensive ore-recovery and a higher price of metal, may be compared to the value included in other forms of commercial enterprise for "good-will." Such forms of enterprise are valued on a basis of the amount which will replace the net assets plus (or minus) an amount for "good-will," that is, the earning capacity. This good-will is a speculation of varying risk depending on the character of the enterprise. For natural monopolies, like some railways and waterworks, the risk is less and for shoe factories more. Even natural monopolies are subject to the risks of antagonistic legislation and industrial storms. But, eliminating this class of enterprise, the speculative value of a good-will involves a greater risk than prospective value in mines, if properly measured; because the dangers of competition and industrial storms do not enter to such a degree, nor is the future so dependent upon the human genius of the founder or manager. Mining has reached such a stage of development as a science that management proceeds upon comparatively well-known lines. It is subject to known checks through the opportunity of comparisons by which efficiency can be determined in a manner more open for the investor to learn than in any other form of industry. While in mining an estimate of a certain minimum of extension in depth, as indicated by collateral factors, may occasionally fall short, it will, in nine cases out of ten, be exceeded. If investment in mines be spread over ten cases, similarly valued as to minimum of extension, the risk has been virtually eliminated. The industry, if reduced to the above basis for financial guidance, is a more profitable business and is one of less hazards than competitive forms of commercial enterprises.

In view of what has been said before, it may be unnecessary

to refer again to the subject, but the constant reiteration by wiseacres that the weak point in mining investments lies in their short life and possible loss of capital, warrants a repetition that the *A, B, C* of proper investment in mines is to be assured, by the "*A*" value, of a return of the whole or major portion of the capital. The risk of interest and profit may be deferred to the *X, Y* value, and in such case it is on a plane with "good-will." It should be said at once to that class who want large returns on investment without investigation as to merits, or assurance as to the management of the business, that there is no field in this world for the employment of their money at over 4%.

Unfortunately for the reputation of the mining industry, and metal mines especially, the business is often not conducted or valued on lines which have been outlined in these chapters. There is often the desire to sell stocks beyond their value. There is always the possibility that extension in depth will reveal a glorious Eldorado. It occasionally does, and the report echoes round the world for years, together with tributes to the great judgment of the exploiters. The volume of sound allures undue numbers of the venturesome, untrained, and ill-advised public to the business, together with a mob of camp-followers whose objective is to exploit the ignorant by preying on their gambling instincts. Thus a considerable section of metal mining industry is in the hands of these classes, and a cloud of disrepute hangs ever in the horizon.

There has been a great educational campaign in progress during the past few years through the technical training of men for conduct of the industry, by the example of reputable companies in regularly publishing the essential facts upon which the value of their mines is based, and through understandable nontechnical discussion in and by some sections of the financial and general press. The real investor is being educated to distinguish between reputable concerns and the counters of gamesters. Moreover, yearly, men of technical knowledge are taking a stronger and more influential part in mining finance and in the direction of mining and exploration companies. The net result of these forces will be to put mining on a better plane.

CHAPTER XX.

THE CHARACTER, TRAINING, AND OBLIGATIONS OF THE MINING ENGINEERING PROFESSION.

In a discussion of some problems of metal mining from the point of view of the direction of mining operations it may not be amiss to discuss the character of the mining engineering profession in its bearings on training and practice, and its relations to the public.

The most dominant characteristic of the mining engineering profession is the vast preponderance of the commercial over the technical in the daily work of the engineer. For years a gradual evolution has been in progress altering the larger demands on this branch of the engineering profession from advisory to executive work. The mining engineer is no longer the technician who concocts reports and blue prints. It is demanded of him that he devise the finance, construct and manage the works which he advises. The demands of such executive work are largely commercial; although the commercial experience and executive ability thus become one pier in the foundation of training, the bridge no less requires two piers, and the second is based on technical knowledge. Far from being deprecated, these commercial phases cannot be too strongly emphasized. On the other hand, I am far from contending that our vocation is a business rather than a profession.

For many years after the dawn of modern engineering, the members of our profession were men who rose through the ranks of workmen, and as a result, we are to this day in the public mind a sort of superior artisan, for to many the engine-driver is equally an engineer with the designer of the engine, yet their real relation is but as the hand to the brain. At a later period the recruits entered by apprenticeship to those men who had established their intellectual superiority to their fellow-workers.

These men were nearly always employed in an advisory way—subjective to the executive head.

During the last few decades, the advance of science and the complication of industry have demanded a wholly broader basis of scientific and general training for its leaders. Executive heads are demanded who have technical training. This has resulted in the establishment of special technical colleges, and compelled a place for engineering in the great universities. The high intelligence demanded by the vocation itself, and the revolution in training caused by the strengthening of its foundations in general education, has finally, beyond all question, raised the work of application of science to industry to the dignity of a profession on a par with the law, medicine, and science. It demands of its members equally high mental attainments,—and a more rigorous training and experience. Despite all this, industry is conducted for commercial purposes, and leaves no room for the haughty intellectual superiority assumed by some professions over business callings.

There is now demanded of the mining specialist a wide knowledge of certain branches of civil, mechanical, electrical, and chemical engineering, geology, economics, the humanities, and what not; and in addition to all this, engineering sense, executive ability, business experience, and financial insight. Engineering sense is that fine blend of honesty, ingenuity, and intuition which is a mental endowment apart from knowledge and experience. Its possession is the test of the real engineer. It distinguishes engineering as a profession from engineering as a trade. It is this sense that elevates the possessor to the profession which is, of all others, the most difficult and the most comprehensive. Financial insight can only come by experience in the commercial world. Likewise must come the experience in technical work which gives balance to theoretical training. Executive ability is that capacity to co-ordinate and command the best results from other men,—it is a natural endowment. which can be cultivated only in actual use.

The practice of mine engineering being so large a mixture of business, it follows that the whole of the training of this

profession cannot be had in schools and universities. The commercial and executive side of the work cannot be taught; it must
be absorbed by actual participation in the industry. Nor is it impossible to rise to great eminence in the profession without
university training, as witness some of our greatest engineers.
The university can do much; it can give a broad basis of knowledge and mental training, and can inculcate moral feeling,
which entitles men to lead their fellows. It can teach the technical fundamentals of the multifold sciences which the engineer
should know and must apply. But after the university must come
a schooling in men and things equally thorough and more ar-
duous.

In this predominating demand for commercial qualifications
over the technical ones, the mining profession has differentiated
to a great degree from its brother engineering branches. That this
is true will be most apparent if we examine the course through
which engineering projects march, and the demands of each
stage on their road to completion.

The life of all engineering projects in a general way may be
divided into five phases:[1]—

1. Determination of the value of the project.
2. Determination of the method of attack.
3. The detailed delineation of method, means, and tools.
4. The execution of the works.
5. The operation of the completed works.

These various stages of the resolution of an engineering
project require in each more or less of every quality of intellect,
training, and character. At the different stages, certain of these
qualities are in predominant demand: in the first stage, financial
insight; in the second, "engineering sense"; in the third, training
and experience; in the fourth and fifth, executive ability.

A certain amount of compass over the project during the

---

[1] These phases do not necessarily proceed step by step. For an expanding
works especially, all of them may be in process at the same time, but if each
item be considered to itself, this is the usual progress, or should be when
properly engineered.

whole five stages is required by all branches of the engineering profession,—harbor, canal, railway, waterworks, bridge, mechanical, electrical, etc.; but in none of them so completely and in such constant combination is this demanded as in mining.

The determination of the commercial value of projects is a greater section of the mining engineer's occupation than of the other engineering branches. Mines are operated only to earn immediate profits. No question of public utility enters, so that all mining projects have by this necessity to be from the first weighed from a profit point of view alone. The determination of this question is one which demands such an amount of technical knowledge and experience that those who are not experts cannot enter the field,—therefore the service of the engineer is always demanded in their satisfactory solution. Moreover, unlike most other engineering projects, mines have a faculty of changing owners several times during their career, so that every one has to survive a periodic revaluation. From the other branches of engineering, the electrical engineer is the most often called upon to weigh the probabilities of financial success of the enterprise, but usually his presence in this capacity is called upon only at the initial stage, for electrical enterprises seldom change hands. The mechanical and chemical branches are usually called upon for purely technical service on the demand of the operator, who decides the financial problems for himself, or upon works forming but units in undertakings where the opinion on the financial advisability is compassed by some other branch of the engineering profession. The other engineering branches, even less often, are called in for financial advice, and in those branches involving works of public utility the profit-and-loss phase scarcely enters at all.

Given that the project has been determined upon, and that the enterprise has entered upon the second stage, that of determination of method of attack, the immediate commercial result limits the mining engineer's every plan and design to a greater degree than it does the other engineering specialists. The question of capital and profit dogs his every footstep, for all mines are ephemeral; the life of any given mine is short.

Metal mines have indeed the shortest lives of any. While some exceptional ones may produce through one generation, under the stress of modern methods a much larger proportion extend only over a decade or two. But of more pertinent force is the fact that as the certain life of a metal mine can be positively known in most cases but a short period beyond the actual time required to exhaust the ore in sight, not even a decade of life to the enterprise is available for the estimates of the mining engineer. Mining works are of no value when the mine is exhausted; the capital invested must be recovered in very short periods, and therefore all mining works must be of the most temporary character that will answer. The mining engineer cannot erect a works that will last as long as possible; it is to last as long as the mine only, and, in laying it out, forefront in his mind must be the question, Can its cost be redeemed in the period of use of which I am certain it will find employment? If not, will some cheaper device, which gives less efficiency, do? The harbor engineer, the railway engineer, the mechanical engineer, build as solidly as they can, for the demand for the work will exist till after their materials are worn out, however soundly they construct.

Our engineer cousins can, in a greater degree by study and investigation, marshal in advance the factors with which they have to deal. The mining engineer's works, on the other hand, depend at all times on many elements which, from the nature of things, must remain unknown. No mine is laid bare to study and resolve in advance. We have to deal with conditions buried in the earth. Especially in metal mines we cannot know, when our works are initiated, what the size, mineralization, or surroundings of the ore-bodies will be. We must plunge into them and learn,—and repent. Not only is the useful life of our mining works indeterminate, but the very character of them is uncertain in advance. All our works must be in a way doubly tentative, for they are subject to constant alterations as they proceed.

Not only does this apply to our initial plans, but to our daily amendment of them as we proceed into the unknown. Mining engineering is, therefore, never ended with the initial determination of a method. It is called upon daily to replan and reconceive, coincidentally with the daily progress of the constructions and operation. Weary with disappointment in his wisest conception, many a mining engineer looks jealously upon his happier engineering cousin, who, when he designs a bridge, can know its size, its strains, and its cost, and can wash his hands of it finally when the contractor steps in to its construction. And, above all, it is no concern of his whether it will pay. Did he start to build a bridge over a water, the width or depth or bottom of which he could not know in advance, and require to get its cost back in ten years, with a profit, his would be a task of similar harassments.

As said before, it is becoming more general every year to employ the mining engineer as the executive head in the operation of mining engineering projects, that is, in the fourth and fifth stages of the enterprise. He is becoming the foreman, manager, and president of the company, or as it may be contended by some, the executive head is coming to have technical qualifications. Either way, in no branch of enterprise founded on engineering is the operative head of necessity so much a technical director. Not only is this caused by the necessity of executive knowledge before valuations can be properly done, but the incorporation of the executive work with the technical has been brought about by several other forces. We have a type of works which, by reason of the new conditions and constant revisions which arise from pushing into the unknown coincidentally with operating, demands an intimate continuous daily employment of engineering sense and design through the whole history of the enterprise. These works are of themselves of a character which requires a constant vigilant eye on financial outcome. The advances in metallurgy, and the decreased cost of production by larger capacities, require yearly larger, more complicated, and more costly plants. Thus, larger and larger capitals are required, and enterprise is passing from the hands of the individual to the financially stronger corporation. This altered position as to the works and finance has made keener demands, both technically and in an administrative way, for the highly trained man. In the early stages of American mining, with the moderate demand on capital and the simpler forms of engineering involved, mining was largely a matter of individual enterprise and ownership. These owners were men to whom experience had brought some of the needful technical qualifications. They usually held the reins of business management in their own hands and employed the engineer subjectively, when they employed him at all. They were also, as a rule, distinguished by their contempt for university-trained engineers.

The gradually increasing employment of the engineer as combined executive and technical head, was largely of American development. Many English and European mines still maintain the two separate bureaus, the technical and the financial. Such organization is open to much objection from the point of view of the owner's interests, and still more from that of the engineer. In such an organization the latter is always subordinate to the financial control,—hence the least paid and least respected. When two bureaus exist, the technical lacks that balance of commercial purpose which it should have. The ambition of the theoretical engineer, divorced from commercial result, is complete technical nicety of works and low production costs without the regard for capital outlay which the commercial experience and temporary character of mining constructions demand. On the other hand, the purely financial bureau usually begrudges the capital outlay which sound engineering may warrant. The result is an administration that is not comparable to the single head with both qualifications and an even balance in both spheres. In America, we still have a relic of this form of administration in the consulting mining engineer, but barring his functions as a valuer of mines, he is disappearing in connection with the industry, in favor of the manager, or the president of the company, who has administrative control. The mining engineer's field of employment is therefore not only wider by this general inclusion of administrative work, but one of more responsibility. While he must conduct all five phases of engineering projects coincidentally, the other branches of the profession are more or less confined to one phase or another. They can draw sharper limitations of their engagements or specialization and confine themselves to more purely technical work. The civil engineer may construct railway or harbor works; the mechanical engineer may design and build engines; the naval architect may build ships; but given that he designed to do the work in the most effectual manner, it is no concern of his whether they subsequently earn dividends. He does not have to operate them, to find the income, to feed the mill, or sell the product. The profit and loss does not hound his footsteps after his construction is complete.

Although it is desirable to emphasize the commercial side of the practice of the mining engineer's profession, there are other sides of no less moment. There is the right of every red-blooded man to be assured that his work will be a daily satisfaction to himself; that it is a work which is contributing to the welfare and advance of his country; and that it will build for him a position of dignity and consequence among his fellows.

There are the moral and public obligations upon the profession. There are to-day the demands upon the engineers which are the demands upon their positions as leaders of a great industry. In an industry that lends itself so much to speculation and chicanery, there is the duty of every engineer to diminish the opportunity of the vulture so far as is possible. Where he can enter these lists has been suggested in the previous pages. Further than to the "investor" in mines, he has a duty to his brothers in the profession. In no profession does competition enter so obscurely, nor in no other are men of a profession thrown into such terms of intimacy in professional work. From these causes there has arisen a freedom of disclosure of technical results and a comradery of members greater than that in any other profession. No profession is so subject to the capriciousness of fortune, and he whose position is assured to-day is not assured to-morrow unless it be coupled with a consideration of those members not so fortunate. Especially is there an obligation to the younger members that they may have opportunity of training and a right start in the work.

The very essence of the profession is that it calls upon its members to direct men. They are the officers in the great industrial army. From the nature of things, metal mines do not, like our cities and settlements, lie in those regions covered deep in rich soils. Our mines must be found in the mountains and deserts where rocks are exposed to search. Thus they lie away from the centers of comfort and culture,—they are the outposts of civilization. The engineer is an officer on outpost duty, and in these places he is the camp leader. By his position as a leader in the community he has a chieftainship that carries a responsibility besides mere mine management. His is the responsibility of example in fair dealing and good government in the community.

In but few of its greatest works does the personality of its real creator reach the ears of the world; the real engineer does not advertise himself. But the engineering profession generally rises yearly in dignity and importance as the rest of the world learns more of where the real brains of industrial progress are. The time will come when people will ask, not who paid for a thing, but who built it.

To the engineer falls the work of creating from the dry bones of scientific fact the living body of industry. It is he whose intellect and direction bring to the world the comforts and necessities of daily need. Unlike the doctor, his is not the constant struggle to save the weak. Unlike the soldier, destruction is not his prime function. Unlike the lawyer, quarrels are not his daily bread. Engineering is the profession of creation and of construction, of stimulation of human effort and accomplishment.

INDEX.

Also from Benediction Books …

Wandering Between Two Worlds: Essays on Faith and Art
Anita Mathias
Benediction Books, 2007
152 pages
ISBN: 0955373700

Available from www.amazon.com, www.amazon.co.uk

In these wide-ranging lyrical essays, Anita Mathias writes, in lush, lovely prose, of her naughty Catholic childhood in Jamshedpur, India; her large, eccentric family in Mangalore, a sea-coast town converted by the Portuguese in the sixteenth century; her rebellion and atheism as a teenager in her Himalayan boarding school, run by German missionary nuns, St. Mary's Convent, Nainital; and her abrupt religious conversion after which she entered Mother Teresa's convent in Calcutta as a novice. Later rich, elegant essays explore the dualities of her life as a writer, mother, and Christian in the United States-- Domesticity and Art, Writing and Prayer, and the experience of being "an alien and stranger" as an immigrant in America, sensing the need for roots.

About the Author

Anita Mathias is the author of *Wandering Between Two Worlds: Essays on Faith and Art*. She has a B.A. and M.A. in English from Somerville College, Oxford University, and an M.A. in Creative Writing from the Ohio State University, USA. Anita won a National Endowment of the Arts fellowship in Creative Nonfiction in 1997. She lives in Oxford, England with her husband, Roy, and her daughters, Zoe and Irene.

Anita's website:
    http://www.anitamathias.com, and
Anita's blog Dreaming Beneath the Spires:
    http://dreamingbeneaththespires.blogspot.com

9 781849 024389

The Church That Had Too Much
Anita Mathias
Benediction Books, 2010
52 pages
ISBN: 9781849026567

Available from www.amazon.com, www.amazon.co.uk

The Church That Had Too Much was very well-intentioned. She
wanted to love God, she wanted to love people, but she was both ham-
pered by her muchness and the abundance of her possessions, and
beset by ambition, power struggles and snobbery. Read about the sur-
prising way The Church That Had Too Much began to resolve her
problems in this deceptively simple and enchanting fable.

About the Author

Anita Mathias is the author of *Wandering Between Two Worlds: Essays on Faith and Art*. She has a B.A. and M.A. in English from
Somerville College, Oxford University, and an M.A. in Creative Writing from the Ohio State University, USA. Anita won a National
Endowment of the Arts fellowship in Creative Nonfiction in 1997.
She lives in Oxford, England with her husband, Roy, and her daughters, Zoe and Irene.

Anita's website:
    http://www.anitamathias.com, and
Anita's blog Dreaming Beneath the Spires:
    http://dreamingbeneaththespires.blogspot.com